T0202150

The Dharma in DNA

The Dharma in DNA

Insights at the Intersection of Biology and Buddhism

DEE DENVER

OXFORD
UNIVERSITY PRESS

OXFORD
UNIVERSITY PRESS

Oxford University Press is a department of the University of Oxford. It furthers
the University's objective of excellence in research, scholarship, and education
by publishing worldwide. Oxford is a registered trade mark of Oxford University
Press in the UK and certain other countries.

Published in the United States of America by Oxford University Press
198 Madison Avenue, New York, NY 10016, United States of America.

Library of Congress Cataloging-in-Publication Data
Names: Denver, Dee, 1973- author.
Title: The Dharma in DNA : insights at the intersection of biology and
Buddhism / Dee Denver.
Description: 1. | New York : Oxford University Press, 2022. | Includes
bibliographical references and index. | Contents: 1. Water—2. Trees
—3. Truths—4. Intersections I—5. Intersections II—6. Sciences
—7. Molecules—8. Identities—9. Bodhi—10. Intimacy—Glossary.
Identifiers: LCCN 2021050429 | ISBN 9780197604588 (hardback) |
ISBN 9780197604601 (epub) | ISBN 9780197604595 (pdf) | ISBN 9780197604618
Subjects: LCSH: Buddhism and science. | Biology—Religious
aspects—Buddhism. | Buddhism.
Classification: LCC BQ4570.S3 D38 2022 | DDC 294.3/3657—dc23/eng/20211202
LC record available at https://lccn.loc.gov/2021050429

DOI: 10.1093/oso/9780197604588.001.0001

1 3 5 7 9 8 6 4 2

Printed by Sheridan Books, Inc., United States of America

To Olivia Belle Hoover

Ko au te awa, ko te awa ko au.
I am the river, the river is me.

—traditional saying of the Whanganui iwi

Contents

Acknowledgments

This book resulted from the love, support, wisdom, and kindness offered to me by countless family members, colleagues and coworkers, teachers, students, and friends over the last decade.

Gratitude goes to Stephanie Swenson, Amani Swenson Denver, and Hirut Swenson Denver for supporting me, putting up with my ever-distracted mind, and keeping the family functioning during my writing retreats. More appreciation goes out to my nuclear family for the many conversations we have had about this book, and for their willingness to allow me to share our family's experiences in it. Thank you. I love you.

More gratitude extends to other corners of my family that include my mother, Kay Denver, and my recently passed father, Phil Denver, Sr., who provided me with the kindness, support, encouragement, and confidence needed to pursue an effort such as this. More thanks go to my brother and sister, Phil Denver, Jr. and Tammy Horn, and to my Aunt Jan Kauffman for their ever-present support and love. Appreciation goes out to my father-in-law, Mike Swenson, for reading and providing feedback on an early draft of the book and for teaching me about aspens. More thanks go to Jan Swenson, Kristen Swenson, and Matthew Mills for their love and support over the years.

Appreciation goes out to my St. Joseph, Missouri high school friends who have shown me companionship, adventure, laughter, and joy throughout my life. Special thanks go out to Colin Mullins, Sona Pai, and Mike Schurke for reading and commenting on an early draft of the book. More gratitude extends to our family friends here in Oregon, and especially the Bland family and the McQuillan family; thanks to Lon and Lex McQuillan for letting me stay in their Waldport beach house for two writing retreats.

This book would not have been possible without two sabbatical leaves granted by Oregon State University. The first sabbatical, during 2012–2013, allowed me to learn about Tibetan Buddhist philosophy and epistemology at Maitripa College in Portland. The second, during 2019–2020, allowed me to travel to Bodhgaya, India, to perform research and compose the final two chapters of the book. Gratitude goes to the OSU Spring Creek Project, which granted me a 2016 writing retreat in the Shoutpouch Creek cabin, hidden away in the Oregon coastal range mountains. Special thanks go out to Brian Atkinson, now at the University of Kansas, for co-instructing an experimental Buddhism and science course with me at OSU in 2016. Gratitude goes to Nana Osei-Kofi, Bradley Boovey, and my fellow 2017 Difference, Power, and Discrimination Academy cohort members for helping me better understand and teach about the nature of power and oppression. I thank my many colleagues and friends in the OSU Department of Integrative Biology, College of Science, and those with the Contemplative Studies Initiative for their engagement and support.

Big thanks go out to the members of my evolutionary genetics research team as it has evolved over the years. Dana Howe, the most skilled scientist I have ever encountered, has been a central and essential source of friendship, support, and wisdom as I have navigated this effort. Eight graduate trainees—Michael Raboin, Emily Bellis, Riana Wernick, Danielle Tom, Ian Morelan, Anh Ha, Sushovita Pal, and Lex Anderson—provided important honest feedback on various versions of this book and its ideas over the last decade. More gratitude goes out to a ninth graduate trainee, Sulochana Wasala, and her wonderful family for hosting me during a 2017 trip to Sri Lanka, where I learned about Theravada Buddhism. Thanks extend to Amila Liyanage for insightful conversations and his willingness to share remarkable photos of Bodhi trees. Two postdoctoral scholars, Amanda Brown and Katie Clark, along with many dozens of undergraduate trainees over the years, provided important insights that helped make this book what it is.

Gratitude goes out to the many teachers and mentors I have encountered and learned from in the world of Buddhism. These include Yangsi Rinpoche, Jim Blumenthal, Bhikkuni Kusuma, Venerable Sarana, Venerable Yifa, Abby Mushin Terris, and the Dalai Lama. Gratitude also goes out to the many teachers and mentors I have encountered and learned from in the world of science. These include Kelley Thomas, Michael Lynch, David Lambert, Lindsey Hutt-Fletcher, and Barbara Taylor. Appreciation extends to my past and present collaborators in biological research that include Charles Baer, Suzanne Estes, Rory Mc Donnell, and Inga Zasada.

This book would not have been possible without wise advice coming from David Barash at the University of Washington and Scott Gilbert, Swarthmore College and University of Helsinki Professor Emeritus. Professor Barash reviewed an early version of the manuscript and has been an important source of advice for academic book writing; Professor Gilbert self-identified to me as a reviewer with Oxford University Press and offered many important suggestions and insights in his review and further along the way. Thanks go out to a second OUP reviewer who chose to remain anonymous.

Gratitude goes out to Jeremy Lewis, senior editor for Oxford University Press, for supporting me and this book effort during the COVID-19 pandemic. Thanks to the Oxford University Press and Newgen Knowledge Works copyediting team for sharing their superb skills with this book.

This work was supported by two Ashoka grants from the Khyentse Foundation, in addition to a John Templeton Foundation Ideas Challenge Award.

1

Water

The rain became the tree. Water molecules in the air condensed into droplets, falling onto the heart-shaped leaves. The drops danced their way down long-tapered leaf tips, descending to join more liquid pooling in the soil below. The water transformed into a tiny trickle that careened between particles of earth and then encountered fine root fibers connected to an underground network of larger roots. The thirsty threads absorbed the water.

The tree became the rain. The water entered the roots and then coursed through the circulatory vessels in the tree, up and away from the roots hidden underground. The water flowed through the trunk and the branches, and then arrived in the leaves. Pores in the leaves opened, and the liquid transformed into gas. The water molecules were now again part of the air, temporarily, before condensing into falling droplets once again.

A wandering thinker, garbed in worn robes, approached the tree. Under the tree he sat, silent and still in meditation, for many days. During that time, he came to see and understand that the rain was the tree, and that the tree was the rain. The wanderer could now see that he was the tree, that he was the rain. The tree became known as the Bodhi tree: the tree of wisdom. Before this moment of awakening, the thinker went by the name Siddhartha Gautama. After his time under the tree, the meditating man became known as the Buddha: the one who woke up.

After the Buddha's enlightenment in ~500–400 BCE, he became a teacher. His teachings at first extended to just a few companions, then grew to include thousands living on the Indian subcontinent during his postenlightenment years, and ultimately spread to many millions of followers across the world in the centuries that followed his death. The Buddha and his teachings shaped numerous civilizations and empires across Asia over the last two millennia. However, with only a few exceptions,[1] the teaching of the Buddha and other Buddhists remained mostly foreign and unfamiliar to the Western world, and to Western science. That is starting to change.

* * *

In the early 2000s, I was a scientist . . . quite the scientist. My young career was on fire. In the preceding years, my dissertation research provided unique new insights into the process of mutation—the speed and patterns by which DNA changes over time. My work applied a uniquely direct approach to examining mutation events by combining laboratory evolution experiments with DNA sequencing technologies. The research revealed a remarkably high rate of mutational change for DNA from one generation to the next, more than ten times faster than previous estimates found in biology and ge-netics textbooks. The first dissertation chapter was published in *Science*,[2] a highly selective and prestigious journal widely considered one of the top two scientific publication venues in the world. The subsequent chapters were also published in top-tier journals, including one in *Nature*[3]—the other member of the "top two" scientific journals. Many scientists aspire to publish in either *Science* or *Nature* once during their entire careers; I published in both before the age of thirty. My precious name appeared first in the author list of these prominent publications; I was obsessed with making sure I got the credit and attention for the work. A steady diet of coffee, hard work, and a hypercom-petitive attitude fueled my scientific fervor. I was motivated by my ego and the joy of proving others wrong.

My growing scientific fame expanded beyond the technical scientific lit-erature. I presented my research to crowded auditoriums at international ev-olution and genetics meetings, and skillfully fielded questions from senior eminent scientists. My discovery of high DNA mutation rates was featured in an *All Things Considered* feature on National Public Radio[4] and made its way into the textbooks, replacing the previous mutation sections that relied on more indirect information. All kinds of people were impressed with me: mentors, collaborators, competitors, family, and friends. However, no one was more impressed with me than my "self."

I lived in Bloomington, Indiana, from 2002 to 2005. During these years, I was a postdoctoral scholar at Indiana University (IU), supported by a pres-tigious fellowship from the National Institutes of Health. My mentor was Michael Lynch, a professor in the IU Department of Biology. Lynch was a renowned and ascending theoretical evolutionary geneticist who was known for his grimly fascinating mutational meltdown theory.[5] During my time at IU, Lynch was developing and gathering support for a new hypothesis on the origins and evolution of genome complexity, which would come to shape ev-olutionary dialogue and research across the world for the coming decades.[6] The Lynch lab was a dynamic and intellectually stimulating environment, with numerous postdocs and graduate students investigating mutation and evolution in diverse animal species, including *Danio rerio* (zebrafish), *Daphnia pulex* (water fleas), and *Caenorhabditis elegans* (microscopic worms). I studied DNA in the tiny worms. The lab was also a highly competitive environment: trainees frequently engaged in spirited debates during lab meetings that on rare occasion degenerated into emotional shouting matches. There were even a few incidents of scientific sabotage. Beyond the Lynch lab, IU Biology was renowned as one of the top departments for evolutionary thinking, and the broader IU and Bloomington community offered a wonderful cultural oasis in southern Indiana.

In 2003, at the peak of my early scientific success, I discovered the Buddha's teachings. I wasn't looking for them. That version of me maintained a purely atheistic outlook, reflective of my Richard Dawkins reading diet and typical of evolutionary science–minded academics. The books of Dr. Dawkins, such as *The Selfish Gene*[7] and *The Extended Phenotype*,[8] occupied my nightstand alongside works by other eminent evolutionists, such as Stephen Jay Gould, and cerebral science fiction writers such as Philip K. Dick. Dawkins's words occupied my mind. His writing appealed to my secular sensibilities and shaped my scientific viewpoints on the central importance of DNA in biology. I was a steadfast believer in the unique power of science as the singular, powerful portal to the truth. This belief offered me comfort, as well as confidence in my "self." Tenzin Gyatso, a Tibetan Buddhist monk better known as His Holiness the 14th Dalai Lama, changed all of this and shifted my sails toward unexpected and unfamiliar waters.

The Dalai Lama's eldest brother, Thubten Jigme Norbu, escaped Tibet from the Chinese invasion in 1950, nine years before the Dalai Lama famously fled across the Himalayas with help from the CIA's Special Activities Division. The Dalai Lama settled in Dharamshala, India, to establish the Tibetan Government in Exile and become an international superstar of Buddhism. The famous lama's brother Norbu settled in Bloomington, Indiana, and became Professor of Tibetan Studies at IU. Norbu, a high lama during his previous life in Tibet, gave up his monastic vows to marry and have three children. In 1979, he established the Tibetan Cultural Center in southwest Bloomington. Throughout his years in Indiana, Norbu was an ardent advocate for his country and people, helping to publicize human rights abuses in Tibet, campaigning for his country's independence from China, and educating local students about Tibetan culture and traditions. Norbu cofounded the International Tibet Independence Movement. He led three long walks for Tibet's independence: an eighty-mile and week-long walk from Bloomington to Indianapolis in 1995; a three-hundred-mile and forty-five-day walk from Washington, DC, to the United Nations Headquarters in New York City in 1996; and a six-hundred-mile and three-month walk from Toronto to New York City in 1997. Norbu died at the age of eighty-six in 2008; his body was cremated on the grounds of the Tibetan Cultural Center in Bloomington.

Before Norbu's death, the Dalai Lama visited his brother in Bloomington on five separate occasions. I was in attendance during the fourth visit, in 2003, when the world-famous Buddhist monk was an honorary guest at an interfaith dedication ceremony for the newly constructed Kumbum Chamtse Ling Temple, built on the Tibetan Cultural Center grounds. My wife, Stephanie Swenson, and I purchased tickets to the event on a whim; they were being sold at the local Tibetan restaurant where we were having lunch one day. We overheard people at a nearby table discussing that Muhammad Ali was going to be there, too, along with some other notable names. We purchased tickets at the counter on our way out the door. I had heard of the Dalai Lama, but I didn't really know much about him other than the fact that he was a famous Buddhist monk who won a Nobel Prize.

We arrived at the Tibetan Cultural Center in our black, geek-chic Volvo station wagon on September 7, 2003. Steph parked in the grass, just off the street. We exited the car and merged into a small but increasing throng of attendees who also came to see and hear words from the Dalai Lama. We walked through a grassy field and then joined a growing line of others who looked like us: a girl in Phish concert T-shirt here, a guy with taped-up glasses wearing a faded button-down shirt and weather-damaged corduroys there. Many hundreds, perhaps thousands of people attended the event. Virtually everyone was white.

We arrived at the Tibetan Cultural Center grounds and passed our tickets to a young hippie girl wearing a volunteer badge. Upon walking through the gate, I started to see very different and unfamiliar kinds of people. Groups of maroon-robed figures, composed mostly of young brown-skinned men with a few elderly white men and women, walked quietly here and there. They all had shaved heads. Some sat under trees. Walking a bit further, we encountered a dozen or so individuals in strange black-cloth clothing, walking in a silent circle. Some had shaved heads; others did not. They took slow, deliberate steps around a small pile of smooth gray rocks that had been stacked on top of one another. Curiosity, blended with a hint of unfamiliar and uncomfortable anxiety and nascent condescension, entered my mind.

A bell sounded, and then Steph and I navigated our way to our seats. A sea of white folding chairs was arranged under an enormous tent complex, and young volunteers helped guests find their places. Along the way, my mood shifted toward restlessness. We sat in our assigned seats, which were far away from the stage; a thick tent pole occluded our view. Thoughts turned toward my worms and experiments back in the lab. I started to feel annoyed and bored, regretting the decision to come to the event. My mind conjured derisive predictions of the Dalai Lama spouting nonsense about human reincarnation and cheesy advice from one of his self-help books. Steph snapped my mind back to present reality with a reminder that we needed to get dog food on the way home. The program then began.

After an introductory procession of other Buddhist monks, nuns, and representatives from other spiritual traditions, the Dalai Lama arrived and sat down on a large chair with a colorful cushion at his back. His translator sat nearby. The famous lama started with a discussion on the value of interfaith dialogue—the importance of finding common ground among the world's many religions and ways of thinking. His words were choppy and sometimes difficult to understand. The Dalai Lama went back and forth between Tibetan and English, requiring the translator to interrupt periodically to translate or rephrase the lama's words. An internal "ugh" entered my mind, and I almost tuned out. But then the Dalai Lama paused in thought and changed his tone, drawing my attention back. The discussion briefly turned toward the Buddhist view of things. The lama spoke about the impermanent nature of all phenomena. *True . . .* he discussed the interconnected nature of everything in the universe. *OK, maybe . . . sort of . . .* then, with a raised finger, the monk sharply said that things should only be believed if they were tested for truth. The statement stunned me. *Was I hearing him right?* The lama continued to discuss the importance of directly seeing truths. It sounded like . . . science to me, not some hokey Eastern religion that believed in deities, supernatural beings, and human rebirth. The possibility entered my mind, a previous impossibility for my mind, that the Dalai Lama's thinking might actually be rooted in some foreign form of logic.

I tried to listen to the remainder of the Dalai Lama's discourse, which returned to emphasis on the importance of all persons and faiths working together to advance compassion and world peace. My racing internal dialogue, however, overwhelmed the lama's words. A few minutes earlier, the monk had delivered something resembling rationality and reason, but in weird words and unfamiliar phrasing. Could it be possible that this superstar Tibetan monk's strange religion might offer something of actual legitimate value, with parallels to science? What about the nonsense? Human rebirth? Deities and mythological creatures? The lama didn't talk about those features of Buddhism, but I knew they were there. My interest was piqued, but my science mind remained skeptical.

The temple dedication program transitioned to a close with the Dalai Lama receiving gifts from fifteen representatives of other faiths. The famous Buddhist monk and other honored guests then departed in a ceremonial procession. Stephanie and I stood from our seats and then discussed the experience a little bit. My mind was still a tempest. Steph shared that she felt inspired by the talk and was happy that she came. She transitioned to talking about other stuff. I stayed quiet, trying to hide my newfound existential uncertainty, which was morphing into some serious distress. I felt embarrassed and didn't want to talk about it. I suggested that we get going, but Steph wanted to stop by the row of stalls selling Tibetan clothing and tapestries before leaving.

We stopped at a small table containing sundry cloths and wall hangings, staffed by an ethereal white-haired woman wearing a wispy purple dress, along with a young brown-skinned monk in maroon robes. A vivid orange and yellow tapestry with unfamiliar Tibetan letters struck a chord with me. I passed cash toward the young man in maroon robes behind the table, but he directed my hand to the smiling elderly white woman beside him. The monk's dark brown eyes, however, maintained direct contact with my eyes. Those eyes penetrated me. In that moment, the monk's eye contact seemed to convey some kind of deep understanding and connection fused with intense, genuine kindness. I didn't know what was happening and felt unnerved. The woman took the bills out of my hands and flashed me a bow with prayer hands; the monk's eyes unlocked from mine and he turned away.

I walked back to the car with my wife, completely befuddled and struggling to digest the events of the last few minutes and the preceding last few hours. We went to the pet store to get food for the dogs, and then I asked Steph to stop by the local used bookstore on our way home. We arrived, and I nervously wandered down the Religion and Spirituality section, keeping my eyes peeled for friends and colleagues who might question my presence in that aisle. I happened upon a copy of a book by the Dalai Lama, *How to Practice: The Way to a Meaningful Life*,[9] which made its way onto my nightstand that evening, right next to my nightly glass of water. I took a drink and then started reading, the water and the words becoming part of me.

* * *

Everything changed. The stream of my mind and the course of my life shifted, in ways that I continuously struggled to digest and understand, during the years following that pivotal experience in Bloomington. I read more books by the Dalai Lama and other Buddhist monks, nuns, and scholars. I intellectually toiled to reconcile my growing fascination with Buddhism with my intellectual foundations in science. In the decade following 2003, I remained a mostly closeted aficionado of Buddhism, limiting my exposure to nighttime reading. I feared responses from my science colleagues, worried about the prospect of career annihilation if I became known as a spiritual quasi-Buddhist. I wasn't really sure how to categorize myself anymore. Steph was aware, of course, seeing me read every night from the growing series of Buddhism books occupying our bookshelf. She didn't seem to make much of it.

Stephanie and I started discussing family in the mid-2000s. We moved from Indiana to New Zealand in 2005, where I took a position as an Allan Wilson Centre for Molecular Ecology and Evolution postdoctoral fellow at Massey University. For the next year, my scientific focus temporarily shifted from worm DNA to penguin DNA. Then, in 2006, I secured a highly competitive and coveted tenure-track Assistant Professor position in the Pacific Northwest of the United States, at Oregon State University. Now that we were settled somewhere with the potential for long-term stability, Steph and I started talking about what we wanted our family to look like. We held a shared, strong conviction to form a family through adoption. Though we agreed to consider trying to have children of "our own" down the road, adopting children was to be the first priority. Despite my obsession with studying the dynamics of DNA as it passed from one generation to the next in worms and penguins, I was completely unconcerned with passing on my own DNA. Instead, for reasons that I did not understand, I held a deep passion for bringing completely random and unknown children, born from the sperm and eggs of some other biological parents, into my life and caring for them. Steph felt the same way, and that is ultimately what we did. Nonetheless, my mind was frequently plagued by the apparent paradoxical contradiction between my obsessions with DNA at work versus its utter irrelevance in my family life. Eventually, the teachings of the Buddha helped me understand this.

* * *

Many years have passed since the Dalai Lama shocked my senses in Bloomington. In that time, the teachings and practices of the Buddha and Buddhists, and their intersections with the teachings and practices of science, increasingly preoccupied my mind. I investigated Buddhism from diverse angles and often with a skeptical scientific approach, seeking lapses in logic and other evidence that might debunk this foreign set of ideas. Buddhism, however, fought back. The teachings of the Buddha and Buddhists that followed posed formidable intellectual challenges. My understanding of Buddhism evolved and matured in unexpected ways. I increasingly recognized the vastness of my ignorance of Buddhism: a tradition considered a religion by some and a philosophy by others, which evolved and diversified in complex ways across Asia over the last ~2,500 years. There were many times when I thought, "I got it," but then later encountered different explanations and framings of Buddhist ideas that revealed an incomplete or inaccurate nature of my previous interpretations. Another magnetic aspect of Buddhism for me is the fact that the teachings are simultaneously, paradoxically, gentle and harsh. The foundational compassion of the Buddha's teachings only functions in the context of philosophical principles that challenge basic concepts of identity and individuality that underpin Western thinking and Western science. My understanding and fascination with Buddhist thinking, and its paradoxes, continue to evolve.

The ongoing investigation of Buddhism progressed alongside my experiences as a professor of evolutionary genetics and my journey as a white father raising two children, adopted from Ethiopia. These three identities—student of Buddhism, professor of genetics and evolution, and white father of Black children—have interactively molded a unique perspective and motivated me to share insights at their intersection.[10] The chapters that follow aim to accomplish three objectives.

The first goal is to share the experiences and lessons learned by an evolutionary geneticist pursuing knowledge and understanding of Buddhist philosophy and practice. Despite a recent surge of written works at the Buddhism–science interface, most biologists and other science-minded people continue to have little to no awareness of the parallels between Buddhism and science. My experiences with Buddhism eventually moved beyond the nightstand books, and they have been diverse. Encounters included time spent on the Theravada Buddhist island of Sri Lanka, a sabbatical learning from a Tibetan Buddhist monk in Oregon, interactions with Zen and Pure Land Buddhists in China and Hawai'i, and engagement with pilgrims representing all of the world's diverse Buddhist traditions during a visit to the place of the Buddha's enlightenment under the Bodhi tree in Bodhgaya, India. How have my experiences with Buddhists affected my thinking about genetics, evolution, and other biological processes? How have these experiences impacted my life as a father in an adoptive transracial family? How have my perspectives on science and family evolved along the way? Chapters 2 and 3 address these questions and provide essential background on the history and diversity of Buddhist thinking. Chapters 4 and 5 share stories of intersections of Buddhist thinkers with biological thinkers.

The second goal is to transform the Buddha's insights, acquired under the Bodhi tree in Bodhgaya more than two millennia ago, into scientific hypotheses and then to test them. A brief tour through the historical development of science, and its relationship to nonscience activities and pseudoscience, will establish the analysis framework. A formal, hypothesis-guided scientific inquiry will be conducted to assess the validity of central Buddhist propositions about the nature of reality. The Buddha's hypotheses will be evaluated using the DNA molecule as a test subject. Seminal experiments conducted over the last century will be revisited, from the classic experiments that revealed the fundamental structural features of DNA, to recent insights arising in today's postgenomic world. Is our modern scientific understanding of DNA consistent or inconsistent with the Buddha's hypotheses? Chapter 6 presents a scientific framework for the Buddha's propositions. Chapters 7 and 8 analyze those propositions by way of DNA.

The third objective is to propose the Buddha's teachings, his logic of compassion, as a moral and ethical foundation for scientists—a framework to guide the decisions they make, the science they do. This strategy for conducting science requires scientists to acknowledge their roles in the scientific process: their subjective biases, the desires of their egos. Going further, scientists are challenged to recognize the ignorance of ego-motivated hypotheses and to look to compassion as a motivator instead. This approach contrasts with the typical practice of Western science whereby scientists view themselves as separate, impartial, purely objective and "neutral" observers in the process. Scientists, perhaps more so than most other people, operate in their daily work lives without acknowledging the effects of their personal

biases, the impacts of emotions. The teachings of the Buddha, however, suggest that such strategies are unwise and will lead to problems. Where do
scientists get their motivations? How do these motivations affect the hypotheses that scientists pursue, the data they collect, and how it is all interpreted?
Chapter 9 proposes a Buddhist approach to science, Bodhi science, founded
on four qualities of Buddhist wisdom: selflessness, detachment, awareness,
and compassion. Further, Bodhi science offers a preventative approach to
help scientists avoid wading into the deceptive waters of pseudoscience.
Chapter 10 returns to the question of my motivations to form a family
through adoption rather than biological reproduction and offers reflections
on my need to ask this question in the first place.

2

Trees

The Buddha's enlightenment under the Bodhi tree was a pivot point in his life and for the world. The story of his experiences—before, during, and after his time under the tree—has been told countless times and in many different ways. Siddhartha Gautama's life history became an embedded component of Buddhist teachings, memorized and passed on in oral traditions among his contemporary students and then on to their students in the years immediately following the Buddha's death. Verbal transmission continued down the centuries, with hand-printed accounts of the Buddha's life and teachings first appearing as part of the Pali Canon, written down in Sri Lanka during 29 BCE in the language of Pali, approximately four hundred years after the Buddha's death. Different Buddhist traditions developed their own written resources. The first woodblock printing of the Chinese Buddhist Canon appeared during the Song dynasty in 971 CE, and the Tibetan Buddhist Canon underwent a final compilation in the fourteenth century CE. In modern times, the Buddha's story became available in diverse books on history and religion, video documentaries, and online resources in English and many other languages.[1] The life path of Siddhartha Gautama, and his evolution into the Buddha, provides critical context for understanding the nature and content of his teachings and insights under the Bodhi tree.

Translation poses a formidable challenge for English-speaking, Western world–thinking students of Buddhism. The Buddha spoke in a language called Magadhi Prakrit, and early Buddhist texts were compiled in a variety of Indic languages, including Pali and Sanskrit. Most English-language works about Buddhism retain the original Pali or Sanskrit for certain key words and phrases. These words represent fundamental concepts in Buddhist philosophy and practice, and sometimes have complex meanings that are refractory to simple, single-word translations. An author's Buddhist background generally determines whether it is the Pali or Sanskrit variety chosen for emphasis in a given work. Those from "southern" (also commonly referred to as Theravada) traditions, practiced in places such as Sri Lanka and Thailand, tend to use Pali words. Those following "northern" (also commonly referred to as Mahayana) traditions, such as the many forms of Zen and Tibetan Buddhism, tend to choose Sanskrit. Sanskrit and Pali words, written in English, generally look similar to one another (or are identical) and often have parallel meanings. However, sometimes cognate Pali and Sanskrit words have distinct connotations, specific to the Buddhist traditions with which they associate.

My investigation of Buddhist philosophy and practice over the last fifteen years was intentionally diverse, studying multiple lineages that evolved across different parts of Asia over the last two and a half millennia. Tibetan Buddhism played an early and prominent role in shaping my analysis, including my initial encounter with the Dalai Lama in Indiana and a 2012–2013 sabbatical at a Tibetan Buddhist college in Portland, Oregon. A journey to Sri Lanka in 2017 provided an opportunity to directly experience a Theravada Buddhist culture; English translations of texts from Sri Lanka's Pali Canon have served as my main primary resource for the Buddha's teachings. Exposure to Zen and Pure Land traditions came with research on Buddhism's evolution on the islands of Kauai, Maui, and Oahu during 2018–2019. A 2020 trip to Bodhgaya, India, the place of the Buddha's enlightenment, exposed me to diverse Buddhist pilgrims and traditions ranging from Myanmar to Taiwan to Tibet to Vietnam.

Given my research approach, both Pali and Sanskrit will be deployed in this book. Sections of the text that focus on Theravada Buddhist thinkers and concepts will emphasize Pali versions of key Buddhist terms, though the Sanskrit analogs will always be provided upon first introduction. For example, subsequent sections of the current chapter will focus on the Buddha's life and teachings derived from Theravada's Pali Canon, so Pali will be emphasized in those passages. Sanskrit will be emphasized in sections focused on Mahayana forms of Buddhism, with Pali versions also provided upon first use. When the text refers to Buddhism in general, with no specific reference to any particular tradition, default emphasis will go to Sanskrit in recognition of my initial exposure to the Buddha's teachings through Mahayana traditions. For example, the book title is *The Dharma* (Sanskrit) *in DNA* rather than *The Dhamma* (Pali) *in DNA*.

* * *

Who was Siddhartha Gautama? Our knowledge and understanding of this man's life derived from ancient *sutras* (Sanskrit; *suttas* in Pali), which are written accounts of the experiences and teachings of the Buddha and his followers. The biographical details of the Buddha are not the primary concern of the *sutras*, though information about different stages of his life is shared in these sacred texts. The *sutras* present two differing identities of the Buddha. He is often simply described as a highly revered, human figure that meditates and offers teachings to his disciplines, nobles, laypersons, and other wandering thinkers encountered at various episodes in his life. Such portrayals present a tone of plausible reality and offer important insights into the historical Buddha. However, some *sutras* offer fantastical descriptions of the Buddha that ascribe to him supernatural powers, such as gaining knowledge of his past lives during his enlightenment experience under the Bodhi tree. Deities, other mystical beings, and supernatural realms also make appearances in some *sutras*. For example, the Buddha's birth is described by Ananda, one of his chief disciples, in the *Acchariya-abbhuta Sutta*[2] of the Pali Canon with a mixture of realistic and miraculous aspects:

> Other women give birth seated or lying down, but not so the mother of the Bodhisatta.[3] The Bodhisatta's mother gave birth to him standing up. (Majjhima Nikaya [MN] 123.15)

> As soon as the Bodhisatta was born, he stood firmly with his feet on the ground; then he took two steps facing north, and with a white parasol held over him, he surveyed each quarter and uttered the words of the Leader of the Herd: "I am the foremost in the world. This is my last birth; now there is no renewal of being for me." (MN 123.20)

A skeptical, scientific reader of *sutras* and other Buddhist texts might be tempted to quickly and completely discount all of Buddhism, given the tradition's recurring supernatural claims and the centrality of human rebirth that are both inconsistent with most scientific outlooks. Such a decision, in my view, would throw a precious wisdom baby out with the supernatural bathwater, the social solvent in which it evolved. The Buddha was born into a time and place where human rebirth and deities were widely accepted, given features of reality. In my research and analysis over the years, I strived to find a middle path between a cautious, critical eye and a patient, open-minded view.

The Path to the Bodhi Tree

Historical scholars agree that Siddhartha Gautama was born sometime around 500 BCE in Lumbini, which is now part of Nepal. He was a privileged prince of Sakya, a small oligarchic state straddling the border of present-day India and Nepal. Thus, many books referred to him as Sakyamuni Buddha ("the enlightened one from Sakya"). Siddhartha experienced a highly sheltered life of leisure and luxury throughout his youth and early adulthood. The Buddha shared some details about his pre-renunciation life during a debate with the wanderer Magandiya, documented in the *Magandiya Sutta*:

> I had three palaces, one for the rainy season, one for the winter, and one for the summer. I lived in the rains' palace for the four months of the rainy season, enjoying myself with musicians, none of whom were men, and I did not go down to the lower palace. (MN 75.10)

Siddhartha lived a life of riches and decadent sensual pleasures. He eventually married and fathered a child. The prince knew nothing other than this sheltered, royal life. One day, with help from his charioteer, Siddhartha sneaked out to discover and experience life outside the palace walls. On this secret expedition, he saw an old person, a sick person, and a corpse . . . all for the first time. These observations shocked the naïve young man to the core, filling him with a deep and unfamiliar existential trauma. These shocking and distressing sights established a preoccupation of human suffering in Siddhartha's mind. Then, the prince encountered another type of unfamiliar person: a wandering ascetic. Homeless, itinerant truth seekers were common in the time and place of the Buddha, persisting off of the charity of others through begging. Some of these drifters traveled alone, others in groups. After meeting with the spiritual wanderer, Siddhartha became inspired to renounce his luxurious life and embark upon the path of truth seeking. The prince discarded his royal garments and replaced them with rags from a rubbish heap. He set out on foot with no possessions, never to return home to his palaces or his family again, as recounted in the *Ariyapariyesana Sutta*:

> Later on, while still young, a black-haired young man endowed with the blessings of youth, in the prime of my life, though my mother and father wished otherwise and wept with tearful faces, I shaved off my hair and

beard, put on the yellow robe, and went forth from the home life into home-lessness. (MN 26.15)

Siddhartha progressed along his truth quest for many years. Along the way, he investigated and learned different forms of meditation, both alone and under the guidance of teachers. Siddhartha trained under two renowned meditation experts of the time, Alara Kalama and Uddaka Ramaputta. He mastered Kalama's approach and then Ramaputta's system, but he became disillusioned and unconvinced that their methods revealed the ultimate truth. Joined by five new companions, the ex-prince next embarked upon a brutal six-year period of starvation and self-abuse, intended to lead him to enlightenment. The experience nearly killed him. Siddhartha eventually rejected this torturous approach of austerities and turned toward a more moderate path to spiritual enlightenment. He transitioned to seek a "middle way," in-between the excessive indulgences of his earlier royal life and the starvation and self-abuse experienced later on as a wanderer. Upon hearing of Siddhartha's new moderate path and seeing him eat some boiled rice and porridge, the five companions became disappointed and disgusted. They saw this new approach as backsliding toward material comforts and luxury, and abandoned him. Siddhartha carried on alone, begging in a nearby village during the mornings and eating one small meal before noon each day. His progress pursuing this new middle path was very swift. One day, Siddhartha came upon a fig tree and sat under it. There he meditated through the af-ternoon and evening and on into the night. This meditation continued for many days, some say for seven weeks, until he dispelled all disturbances of the mind and realized complete enlightenment. Siddhartha is said to have achieved three forms of knowledge as a result of his enlightenment: know-ledge of the details of his past lives, knowledge of the process of the passing away and reappearance of beings, and insight into the Four Noble Truths as the path to enlightenment.

After his experience under the Bodhi tree, Siddhartha became known as the Buddha. After some days debating whether he should keep this new-found wisdom to himself and pass into *nirvana* (Sanskrit; *nibbana* in Pali), a transcendent state where the cycle of birth and death ends and all suffering is quenched, he opted instead to share his insights with others. The Buddha presented his first teaching at the Deer Park in Isipatana (known today as Sarnath, India), to the five former companions that had earlier abandoned him. They accepted and rejoiced in his teachings and became the Buddha's first set of disciples. This event became known as the First Turning of the Wheel of Truth—the first teaching of the Buddha.

The Buddha traveled across the Ganges basin with an ever-increasing contingent of followers, offering teachings on his newfound path to liberation from suffering and insights into the nature of reality and how the universe works. The Buddha's following grew and grew across the years, swelling in number to the thousands. Ever-increasing numbers of men and women heard the Buddha's teachings, with many abandoning their previous lives to shave their heads and become renunciates, followers of this new teacher and his middle path. They gave up all material possessions, except for robes and a begging bowl. They ate food acquired during daily morning alms rounds, but they did not eat after noon. They listened to teachings from the Buddha and followed his meditative instructions so that they, too, could directly experience enlightenment. This community of fellow practitioners following the Buddha's path to enlightenment became known as the *sangha* (Sanskrit; also *sangha* in Pali).[4] The Buddha spent the rest of his life, forty-five years, as a teacher of his newly discovered wisdom.

The Four Noble Truths

The Buddha's teachings are called the *dharma* (Sanskrit; *dhamma* in Pali). This word signifies the truth as transmitted by the verbal teachings of the Buddha and subsequent enlightened persons. Use of the word *dharma* can also extend to conceptual and phenomenological vehicles by which the truth is expressed, necessary to make it comprehensible.

The Buddha was a pragmatist. He saw humanity as being in urgent need of help and considered deep metaphysical questions a waste of time and energy. The Buddha's views on the distracting nature of philosophical musings were expressed in the *Culamalunkya Sutta*, in a passage that has since become widely known as the "Parable of the Poison Arrow." In this *sutta*, a follower of the Buddha named Malunkyaputta became dissatisfied with the Buddha's refusal to make declarations on metaphysical questions such as "Is the world finite or infinite?," which prompted the Buddha to explain his rationale for silence in response to such questions:

Suppose, Malunkyaputta, a man was wounded by an arrow thickly smeared with poison, and his friends and companions, his kinsmen and relatives, brought a surgeon to treat him. The man would say: "I will not let the surgeon pull out this arrow until I know whether the man who wounded me was a noble or a brahmin[5] or a merchant or a worker." And he would say: "I will not let the surgeon pull out this arrow until I know the name and the clan of the man who wounded me; . . . until I know whether the man who wounded me was tall or short or of middle height; . . . until I know whether the man who wounded me was dark or brown or golden skinned; . . . until I know whether the man who wounded me lives in such a village or town or city; . . . until I know whether the bow that wounded me was a longbow or a crossbow; . . . until I know whether the bowstring that wounded me was fiber or reed or sinew or hemp or bark; . . . until I know whether the shaft that wounded me was wild or cultivated; . . . until I know with what kind of feathers the shaft that wounded me was fitted—whether those of a vulture or a heron or a hawk or a peacock or a stork; . . . until I know with what kind of sinew the shaft that wounded me was bound—whether that of an ox or a buffalo or a deer or a monkey; . . . until I know what kind of arrowhead it was that wounded me—whether spiked or razor-tipped or curved or barbed or calf-toothed or lancet shaped." All this would still not be known to that man and meanwhile he would die. (MN 63.5)

The *dharma* discovered by the Buddha under the Bodhi tree was practical in nature, serving as a mechanism to alleviate *duhkha* (Sanskrit; *dukkha* in Pali). The word *duhkha* is often simply translated as "suffering," though the scope of this word in relation to the Buddha's wisdom is much broader. Physical pain and discomfort are included under the *duhkha* umbrella, though the word also indicates a general uneasiness, or dissatisfaction in life. *Duhkha* can manifest as profound, like how one feels after a loved one dies. It comes through deep existential despair and suicidal emotions. More moderate forms of *duhkha* arise through daily anxieties and obsessions. *Duhkha* also appears through trivial matters, like a hangover, or the fly buzzing by the lamp during a night of insomnia. Boredom is a form of *duhkha*. Broadly, *duhkha* derives from the foggy, cyclic world of ignorance in which unenlightened people mindlessly stumble around in blind reactionary states, a place referred to as *samsara* (Sanskrit; also *samsara* in Pali). For most Buddhists, *samsara* is the realm into which people are repeatedly reborn. *Duhkha* can also simply and neutrally refer to the lived experience.

The Buddha's *dharma* was taught as The Four Noble Truths, and the reality of *duhkha* was the first of these truths. *Duhkha* served as a skillful strategy to commence exposition of the *dharma*: everyone experienced pain, suffering, dissatisfaction, and unhappiness in some form or another. People could relate to *duhkha*. Most also presumably had a keen interest in eliminating it, or at least having less of it in their lives. Siddhartha's shocking exposures to sickness, old age, and death on that first venture outside the palace walls catalyzed his own journey toward enlightenment. The Buddha acknowledged that happiness and joy also appeared in life, though such moments were always fleeting and impermanent in nature. *Duhkha* was always right around the corner. Each of the Buddha's Four Noble Truths related to *duhkha*:

1. the truth of *duhkha*
2. the truth of the origin of *duhkha*
3. the truth of the cessation of *duhkha*
4. the truth of the path leading to the cessation of *duhkha*

* * *

Much like a scientist, the Buddha's observations motivated him to seek explanations. The first of the Four Noble Truths identified *duhkha* as a universal feature of the lived experience. The Buddha devised a classificatory framework called the five *skandhas* (Sanskrit; *khandas* in Pali), usually translated as the five aggregates, for describing the lived experience of humans, the mechanism by which they engaged in *duhkha*. The first of these factors was matter, which includes the external material objects as well as an individual's physical body with all of its sense faculties. The second *skandha* was sensation, the initial interaction between subject and object that could be pleasant, painful, or neutral. Perception was the third factor, and it denoted the sensory and mental processes that recognized objects of sensation and gave them labels. The fourth *skandha*, mental formations, encompassed the entire diverse suite of volitional, emotional, and intellectual aspects of mental processes that lead a person to initiate response and action. Consciousness was the fifth and final *skandha*, denoting the basic awareness of a given object of cognition, previously sensed, perceived, and subject to mental formations in the preceding *skandhas*. The first of the Buddha's truths is that suffering exists, and it manifests in the context of the five *skandhas*.

What were the causes of the *duhkha* effect? The second of the Four Noble Truths identified craving and clinging as the origins of *duhkha*. Attachments and aversions to material possessions, people, and various types of mental constructs were identified as sources of suffering. These processes functioned through the five *skandas* framework of human experience. The Buddha frequently highlighted cravings for sensual pleasures as a common cause of *duhkha*. Sariputta, one of the Buddha's foremost disciples in the Pali Canon and Theravada Buddhism,[6] outlined six categories of cravings in a discourse about the Four Noble Truths, presented in the *Sammaditthi Sutta*:

> And what is craving, what is the origin of craving, what is the cessation of craving, what is the way leading to the cessation of craving? There are these six classes of craving: craving for forms, craving for sounds, craving for odors, craving for flavors, craving for tangibles, craving for mind-objects. (MN 9.38)

In the same teaching, Sariputta identified more types of craving and clinging that lead to *duhkha* that included clinging to views, clinging to rules and observances, and clinging to a doctrine of "self." Unpacking these more cryptic varieties of clinging requires an exposition of two key and interdependent principles that underpin all of the Buddha's teachings and constitute unique features that distinguish Buddhism from all of the other major world religions: *anitya* and *anatman*.

Anitya (Sanskrit; *anicca* in Pali), commonly translated as impermanence, was one of these two foundational principles in the Buddha's teachings. All features of the universe were transient and unstable, according to the *dharma*. All phenomena were said to lack any kind of permanent, everlasting existence or characteristics. Rather, everything everywhere was perpetually changing, in a constant state of dynamic flux. At the cosmological scale, the Buddha espoused a view of the universe marked by infinite cycles of expansion and contraction. At intermediate scales, the human experience offered a frequently used example of *anitya* in Buddhist teachings. There was birth, childhood, the teen years, adulthood, parenthood, old age, death, and (for most Buddhists) rebirth to start the cycle of *samsara* all over again. The Buddha explained the impermanent nature of sensations, the second *skandha*, to a skeptical wandering ascetic in the *Dighanakha Sutta*:

> Pleasant feeling, Aggivessana,[7] is impermanent, conditioned, dependently arisen, subject to destruction, vanishing, fading away, and ceasing. Painful feeling too is impermanent, conditioned, dependently arisen, subject to destruction, vanishing, fading away, and ceasing. Neither painful nor pleasant

feeling too is impermanent, conditioned, dependently arisen, subject to de-
struction, vanishing, fading away, and ceasing. (MN 74.11)

On smaller scales, Buddhist meditation and mindfulness practices emphasize awareness of the impermanent nature of the breathing process, air
flowing into and out of the lungs. The breaths arise and pass away in succes-
sion, from one moment to the next. Oxygen molecules are inhaled, followed
by carbon dioxide exhaled in one breath. New molecules are inhaled and
exhaled in the next breath. *Anitya* was a universal truth for the Buddha's
dharma, no exceptions.

Anatman (Sanskrit; *anatta* in Pali), commonly translated as "non-self,"
served as a second foundational and distinctive principle of the Buddha's
teachings. Walpola Rahula, a twentieth-century Theravada Buddhist monk
and scholar from Sri Lanka,[8] chose to translate *anatman* as "no soul" in his
book *What the Buddha Taught*.[9] Contrary to widespread and cherished
beliefs of humans, the Buddha instructed that beings lacked any kind of
inherent, immutable "self" nature to serve as a foundation for individual
identity. The Buddha's *anatman* proposition was radical, a direct counter
to the centrality of the *atman* (Sanskrit; *atta* in Pali), "true self" principle of
the Vedic religions and cultures that prevailed in the time and place of the
Buddha, and were the antecedents to modern-day Hinduism. According to
the Buddha, the "self" concept held only a conventional sort of validity, as
a convenient though misleading shorthand descriptor for human existence.
"Self" implied a sense of stability and permanence for what was in truth a
highly unstable and impermanent situation.

The interwoven trio of *anatman*, *anitya*, and *duhkha* together make
up Buddhism's "three marks of existence." The physical body and mental
factors are impermanent, in a constant state of transition from one impermanent state to the next. A being's existence is also a function of dynamic
contingencies and interdependencies, operating through the five *skandhas*,
with the phenomenological world and one's mental processes. Thus, at any
given slice in time, the "self" that was there a moment ago has vanished,
replaced by a new "self" that instantaneously suffers the same fate, and so
on. Unenlightened beings cling to views, unwilling to set aside erroneous
outlooks when more logical alternatives are presented. They cling to rules
and observances, relying on attachments to the false notions of stability
offered by other misguided teachings. They cling to the delusion of "self" and
dwell in *duhkha*.

The third of the Four Noble Truths revealed the potential for the cessation of *duhkha*. The Buddha realized *nirvana* during his enlightenment experience under the Bodhi tree, and that awakening was available to anyone capable and willing to pursue it. Siddhartha Gautama did not consider himself to be a special case, some kind of individual uniquely endowed with exceptional abilities. Many members of the Buddha's *sangha* are said to have achieved complete enlightenment, following the teachings and meditational methodologies espoused by the Buddha. In the Pali Canon, such individuals, and the Buddha himself, were referred to as *arahants* (Pali; *arhats* in Sanskrit). Sariputta was an *arahant*, and in the *Saccavibhanga Sutta* he described the third truth as the "remainderless fading away and ceasing, the giving up, relinquishing, letting go, and rejecting of craving."[10]

The Buddha's enlightenment was not something that could be accomplished through mere intellectual reasoning alone or by simply listening to teachings of the *dharma*. Rather, complete liberation from *duhkha* required one to directly "see" the true nature of reality through advanced meditation. In *What the Buddha Taught*, Walpola Rahula deployed an apt analogy to describe the futility of trying to explain enlightenment through words:

> . . . there cannot be words to express that experience (nirvana), just as the fish had no words in his vocabulary to express the nature of solid land. The tortoise told his friend the fish that he just returned to the lake after a walk on land. "Of course," the fish said, "you mean swimming." The tortoise tried to explain that one couldn't swim on the land, that it was solid, and that one walked on it. But the fish insisted that there could be nothing like it, that it must be liquid like his lake, with waves, and that one must be able to dive and swim there.

Though *nirvana* itself could not be explained with words, the Buddha formulated the Noble Eightfold Path to describe the route to eliminate craving and clinging, and extinguish *duhkha*. This path constituted the fourth of the Four Noble Truths. The eight spokes supporting the *dharma* wheel of the Noble Eightfold Path were as follows:

1. right view
2. right intention
3. right speech
4. right action

5. right livelihood
6. right effort
7. right mindfulness
8. right concentration

In the *Mahacattarisaka Sutta*,[11] the Buddha explained that the first seven spokes served as supports, requisites for the establishment of the ultimate goal of right concentration. With "right view," a person had an accurate outlook of the world around them, rooted in recognizing the realities of *duhkha*, *anitya*, and *anatman*. Further, knowledge of the Four Noble Truths was required for an individual to have the right view.

With initial establishment of the right view of things, the next step was to cultivate right intentions. The objective to lead life as a renunciate follower of the Buddha's teachings was one form of right intention. According to the Buddha and his students, right intentions involved abstaining from wrong intentions such as ill will, cruelty, and thirst for sensual pleasures. Right intention led to right speech and right action. Under the spoke of right speech, one guided by the right view and right intention abstained from speaking falsehoods and avoided harsh and malicious words. Further, gossip and idle chatter did not pass the lips of one guided by the Buddha's right speech. The spoke of right action required one to refrain from killing living beings, not steal, and abstain from sensual pleasures.

With right livelihood, both renunciate followers of the Buddha and laypersons were expected to carry on in life without scheming, belittling others, and obsessively pursuing gain. In the *Vanijja Sutta*,[12] the Buddha revealed that merchants who follow his teachings were expected to specifically avoid business involving weapons, meat, intoxicants, poison, and human beings. The spoke of right effort inspired one to awaken zeal and pursue the path to enlightenment with an energetic, motivated mind.

Right mindfulness was the final supporting spoke, leading to the final goal of right concentration. The mindfulness concept has developed into a major emphasis of modern-day Western psychological theory and counseling practice, and also made inroads into the curriculum of American schools, from kindergarten to university. Despite common misconception, mindfulness did not simply denote a state of "being present," but rather involved inquiry and reflective analysis. In the *Satipatthana Sutta*, the Buddha identified four foundations of mindfulness: contemplation of the body, contemplation of feeling, contemplation of mind, and contemplation of mind-objects.

Establishment of the four foundations is necessary for the realization of *nirvana*, and it begins with mindfulness of breathing:

> And how, *bhikkhus*,[13] does a *bhikkhu* abide contemplating the body as a body? Here a *bhikkhu*, gone to the forest or to the root of a tree or to an empty hut, sits down; having folded his legs crosswise, set his body erect, and established mindfulness in front of him, ever mindful he breathes in, mindful he breathes out. Breathing in long he understands: "I breathe in long"; or breathing out long he understands: "I breathe out long." Breathing in short he understands: "I breathe in short"; or breathing out short he understands: "I breathe out short." He trains thus: "I shall breathe in experiencing the whole body"; he trains thus: "I shall breathe out experiencing the whole body." (MN 10.4)

Right concentration, a mindfully aware state of equanimity, was the eighth and final spoke of the Noble Eightfold Path, realized upon achieving a synthetic mastery of the seven supporting spokes. Other *suttas*, however, organized This Noble Eightfold Path in fashions that differed from the *Mahacattarisaka Sutta*. For example, the *Culavedalla Sutta*[14] explained that the eight spokes were bundled into three "baskets" of training. Right speech, right action, and right livelihood made up the basket of moral discipline; right effort, right mindfulness, and right concentration composed the basket of concentration; right view and right intention made up the basket of wisdom. The Noble Eightfold Path had been further elaborated upon and integrated into diverse practices and detailed stepwise paths to enlightenment that varied among the many forms of Buddhist practice that have evolved over the last 2,500 years. This path and the broader Four Noble Truths, however, constituted core, foundational features of all forms of Buddhism.

Cause and Effect

The Buddha proposed *pratityasamutpada* (Sanskrit; *paticca samuppada* in Pali) as the cause-and-effect framework for how the universe functions. *Pratityasamupada*, a fundamental principle shared by all forms of Buddhism and described in multiple *sutras* and diverse Buddhist commentaries, is often translated as "dependent arising." Alternative translations include "dependent origination," "conditioned arising," "conditioned genesis," "dependent co-arising," and "mutual interdependent arising." This principle of causality served as the mechanism to explain the origin and cessation of *duhkha* (i.e., the second and third noble truths). *Pratityasamupada* also underpinned the Buddha's concept of *karma* (Sanskrit; *kamma* in Pali): how one's present volitional actions influence one's future state, within and across lifetimes. The Buddha explained a general gist of *pratityasamutpada* to his disciple Ananda in the *Bahudhatuka Sutta*:[15]

> When this exists, that comes to be; with the arising of this, that arises. When this does not exist, that does not come to be; with the cessation of this, that ceases. (MN 115.11)

In Theravada traditions, the principle of dependent origination is often described through a circular set of twelve interdependent steps that describe the process of human life and rebirth, with each one depending on the preceding step in both directions. In *What the Buddha Taught*, the Sri Lankan Theravada monk and scholar Walpola Rahula described *pratityasamutpada* as such:

> On this principle of conditionality, relativity and interdependence, the whole existence and continuity of life and its cessation are explained in a detailed formula which is called *paticca samuppada*.

In other Buddhist words, all *dharmas* (phenomena, things, etc.) arise and then disappear in mutual interdependence of other *dharmas*. Such generalized descriptions of *pratityasamutpada* are common, though there are more specific features of the principle that get lost in terse descriptions. Piyadassi Thera was another prominent twentieth-century Theravada Buddhist monk and scholar from Sri Lanka who served as the lead editor for the Buddhist Publication Society and published an essay, *Dependent Origination* (Paticca Samuppada),[16] which outlined two additional key features of the Buddha's view of cause and effect. First, Piyadassi emphasized that there was never a true, ultimate First Cause: the universe had no beginning and would never have a true ending. The monk wrote of the regress problem that resulted when trying to rationalize a First Cause:

> If one posits a "First Cause," one is justified in asking for the cause of that "First Cause," for nothing can escape the law of condition and cause which is patent in the world to all but those who do not see it.

Second, Piyadassa also negated single causes in the *pratityasamutpada* framework. The Buddha's system of cause and effect considered all *dharmas* as contingent upon countless causes and conditions. The tendency to distill some resultant phenomenon as the effect of some separate singular cause, according to the Buddha, was rooted in ignorance. Piyadassa further explained:

> The doctrine of *paticca-samuppada* can be illustrated by a circle, for it is the cycle of existence, *bhavacakka*. In a circle any given point may be taken as the starting point. Each and every factor of the *paticca samuppada* can be joined together with another of the series, and therefore no single factor can stand by itself or function independently of the rest. All are interdependent and inseparable. Nothing is independent, or isolated. Dependent origination is an unbroken process. In this process nothing is stable or fixed, but all is in a whirl. It is the arising of ever-changing conditions dependent on similar evanescent conditions.

Mahayana views of *pratityasamutpada* were also rooted in the *sutras* found in the Pali Canon but also evolved in lineage-specific ways as the Buddha's teachings took their northern historical route through places such as Tibet, China, and Japan. The metaphor of Indra's net—describing the universe as an infinite jeweled lattice, with mirroring at all intersectional nodes in the net—originated with the Huayan Mahayana tradition that flourished during the Tang dynasty in China, and has become a widely deployed descriptor for *pratityasamutpada*. In *Ocean of Reasoning*,[17] an influential fourteenth- to fifteenth-century Tibetan Buddhist monk named Tsonkhapa tied *pratityasamuptada* to *duhkha*, the portal to the Four Noble Truths:

> Whoever sees dependent arising also sees suffering.

The Path to the Acacia Tree

Four years after my path pivoted toward Buddhism in Bloomington, another sharp turn awaited in Africa. After my postdoctoral training time in Indiana and New Zealand, in 2006 I began duties as an Assistant Professor at Oregon State University. Stephanie secured a position as a middle school math teacher with the Corvallis School District, and we moved into a cute and mossy green house at the top of a hill, surrounded by colossal Douglas fir trees. Just down the street, an idyllic path through the woods led to our local elementary school. I daydreamed about walking children down the street, and into the path through the trees. With the prospect of professional stability within sight, Stephanie and I started the process of becoming parents.

We chose to pursue adoption from Ethiopia. The decision to form our family in this fashion was welcomed and supported by our family, friends, and colleagues. It also sometimes brought some puzzlement and apprehensive questions. When sharing the news with others, initial responses included hearty words of congratulations and encouragement, though often also followed by some version of the question, "Why adopt from Ethiopia?" Both Stephanie and I would usually initially respond, "I don't know." Something unknown drew us both to seek out a life as adoptive parents. Our responses also often shared the variety of diverse circumstantial factors that influenced our path and the extensive homework we had done. There was a nearby adoption agency in Oregon with a long-standing, reputable relationship with an orphanage system in Ethiopia. Further, this Horn-of-Africa nation's international adoption policies were clear and offered expedient timelines. We talked to other nearby international adoptive families and learned about the psychological and social challenges our future children might face, such as attachment disorders and the experience of children of color growing up in Oregon. Our decision was certainly impacted by broader American society in the mid-2000s, when diverse transracial adoptive families were becoming increasingly common and then-Senator Barack Obama's political career was on the rise. The version of me existing back then foresaw an upcoming future of racial harmony and social progress in America. Steph and I were excited about our family vision, and quickly and confidently proceeded through the litany of paperwork, adoption agency home visits, and travel arrangements required to realize our dream.

We decided to pursue simultaneous adoption of two children, reasoning that it made more sense to do all of the travel and paperwork once rather than twice. Stephanie and I also wanted the two children to "have each other." We hoped that the presence of another Black adoptive sibling in each other's lives might offer some kind of comfort as they grew up in the social environment of Corvallis, Oregon: white, liberal academics could be found astride bicycles at virtually every street corner, but Black people and other ethnic minorities were exceedingly rare. The lack of racial and ethnic diversity in Corvallis posed a constant challenge that Steph and I repeatedly discussed and debated as we navigated our path toward forming a transracial adoptive family. We did our best to think through everything we could do to make our Black children-to-be happy and healthy, but we also recognized and lamented the reality that much was out of our control. We second-guessed ourselves a lot but kept moving forward. During the early fall of 2007, we were matched with two Ethiopian infants: a boy named Amanuel and a girl named Hirut.

* * *

In December 2007, Stephanie and I took an Ethiopian Airlines flight from Washington, DC, to Addis Ababa. After the arrival of our preceding connecting flight into the Dulles Airport, we found our way to the international terminal and then our gate; after getting some coffee and snacks, we took seats in the empty waiting area by the gate. We were prepared for the big trip, with checked luggage full of infant clothing gifted to us by our family, friends, and colleagues. An extra suitcase bulged with gifts and supplies for the orphanage that housed many dozens of children that we would not be bringing back to Oregon. My carry-on backpack held important items such as printed photos of Amanuel and Hirut, a laptop filled with precious scientific data, and two books by Stephen Batchelor: *Buddhism Without Beliefs: A Contemporary Guide to Awakening*,[18] and *The Awakening of the West: The Encounter of Buddhism and Western Culture*.[19] Batchelor was an excellent writer and a pioneer in the development of a formative "secular Buddhism" scene in America and Europe. While sitting on the uncomfortable seats of the airport lobby, however, my eyes were focused on the lines and lines of DNA sequence data displayed by my laptop. I disappeared into another world dominated by the strings of A's, C's, G's, and T's appearing on my computer's screen. I was running programs on my laptop that lined up the different stretches of letters in different ways, looking for patterns of similarity and difference in the DNA.

After an hour or so immersed in the lines of DNA sequence information, I smelled a change in my surroundings. A strong scent of cologne emanated from an exceedingly thin young Ethiopian man who slept in the seat right next to me. I had not even noticed that he arrived. He was dressed in a dark brown business suit and a black briefcase rested on his lap. I looked up and then looked around. The previously vacant waiting area had transformed into a crowded sea of Black people. The scene surprised me, and an unfamiliar sense of discomfort crept into my psyche. I kept looking around. Two hijab-wearing grandmothers chatted in the chairs across from me. A group of six or seven children raced around the lobby, shouting taunts to one another in an unfamiliar language that I assumed to be Amharic. There were a few other pairs of white people whom I surmised might be on similar adoption missions. After a few moments reflecting on the source of my discomfort, a sort of unease that I had never felt before, the gate agent announced that boarding was to begin. Stephanie and I joined the growing line, making our way onto the plane that would deliver us to a new reality.

After landing and navigating the unfamiliar complexities of the Addis Ababa airport, Stephanie and I traveled to the orphanage guest house along with our lawyer, who would help us navigate the adoption process. Then, Stephanie and I traveled by van to the orphanage that cared for Amanuel and Hirut. After arriving, we were placed in a jungle-themed receiving room adorned with a big comfortable sofa and children's toys; we were anxious for the big moment to arrive. So many hours had been spent staring at pictures of our children-to-be. My mind wandered, constructing a fantasy of an immediate, intimate connection with the infants who would soon arrive. I envisioned a scenario something like that optic connection with the maroon-robed monk at the temple dedication event in Bloomington, four years earlier. A nurse suddenly walked through the door, snapping me out of my daydream, and delivered a small infant girl into my arms. My shoulders tensed. The nurse said, "Hirut." The infant's enormous eyes were beautiful and danced around the room. Her head was bald, just like in the pictures. A trickle of drool cascaded from her mouth and then down my shoulder, traveling the length of my arm until it hit my thumb. Hirut started to squirm and cry in my clumsy arms. I attempted to appease her with some hip bouncing and a teething ring, but my actions and the toy failed to soothe her distress. We never made eye contact. At the same time, a second nurse delivered Amanuel into Stephanie's arms and said, "Amani." My wife's left hand skillfully supported the boy's small curly-locked head as she held him, and his eyes became transfixed by Stephanie's earring. He reached for the small piece of jewelry in her ear with a serious look on his face, remaining calm in her arms. Stephanie smiled. After a few minutes Stephanie and I sat next to one another on the couch, Hirut in tears and Amani at peace. We exchanged children. Upon arriving in my lap, Amani started to cry.

My early fatherhood flails continued in the coming days, though Amani and Hirut gradually became more and more comfortable with me. Our budding family spent time together at the orphanage guesthouse where we stayed, joined by other white parents and their newly adopted Ethiopian babies. An amazing kitchen crew made marvelous meals and brewed strong dark coffee in the mornings. The cooks loved seeing all of the children, and being with them. Big smiles lit their faces every time we passed by the kitchen with Amani and Hirut, and the motherly women snatched them from our arms. They tickled and fawned over the children for a few minutes before passing them back to us and returning to their work.

One day around the dining room table, the adoptive parents gathered to share a meal, and the topic of baby names arose. We were a group of six, all from the Pacific Northwest of North America and all white. One couple was Mormon and didn't drink coffee, thankfully leaving more for us. The other couple was not religious and adopted a girl who was from Awassa, the same southern region of Ethiopia where Hirut was born. Amani was born in Gondar, in northern Ethiopia. The adoptive parents all got along rather well. One of the mothers sparked a conversation about the joy of choosing a new name for her baby. The other four compared the usefulness of different baby name books and online lists. One couple selected an alternative African name for their baby that was easier to pronounce than the child's original Amharic name; the other couple chose an English name that had been passed on in their family. Stephanie and I remained awkwardly quiet during the discussion. Our silence was eventually noticed, and one of the fathers steered the conversation toward the upcoming adoption court date at the embassy. Just as we wanted Amani and Hirut to have one another as links to their African origins, Stephanie and I decided to keep their names as another lifelong reminder of their Ethiopian roots.

After the meal, Stephanie and I returned to our room in the guest house; Steph decided to read in the room, and I wandered outside to read in the sun. I stopped by our bedroom on the way out to grab the copy of Batchelor's *The Awakening of the West* from my backpack, and then shuffled past the kitchen crew and out the back door to sit under a big acacia tree that shaded the guesthouse yard. Romeo, the small and charismatic guesthouse dog, joined me under the tree. My fingers found the bookmark and opened the book to a page that started with the story of Menander, the Greek king of India, and his dialogue with a Buddhist monk, Nagasena. Centuries before, this story was written down in Pali as a Theravada Buddhist *sutta*, *Milinda Panha*, in the Burmese Buddhist Canon. The story, however, was not included in the Pali Canon in Sri Lanka's version of Theravada Buddhism. The story also appeared in the Mahayana Chinese Buddhist Canon.

After reading the recount of the discourse between Menander and Nagasena in Batchelor's book, I laughed out loud. Romeo tilted his head and gave me a puzzled look. I had believed that the dinner-table discussion of names was over for me, but the Buddhist book brought the topic of names and labels back into my mind. This historical dialogue, known as Menander's Questions and thought to have taken place ~150 BCE, helped me understand the challenging *anatman* concept of Buddhism in a new way. The well-known conversation between King Menander and the monk Nagasena went like this:

"Who are you?" asked Menander.

"I am Nagasena, sir," replied the monk. "As such am I known to my fellow renunciates, but although my parents gave me this name, it is merely a denotation, a designation, an appellation, a conventional usage. Nagasena is only a name since there is no person to be found here."

Menander retorted, "But if no person can be found, who gives you robes and alms? Who is it enjoys them? Who is it that keeps vows, practices meditation, and realizes nirvana? Let us take a closer look . . . Could it be that the hairs on your head are the venerable Nagasena?"

"Oh, no sir," replied Nagasena.

"Then what about the hairs on your body?"

"Impossible, sir."

"How about your nails, teeth, skin, flesh, sinews, bones, marrow, kidneys, heart, liver, membranes, spleen, lungs, intestines, stomach, excrement, bile, phlegm, pus, blood, sweat, fat, tears, saliva, mucus, or urine? Your brain, perhaps?"

"None of them, sir."

"So, if you are not your body, might you then be a feeling of pleasure or pain, a perception, a mental impulse, or a state of consciousness?"

"No, sir."

"If you are neither body nor mind, do you somehow exist apart from your body and mind?"

"Not at all, sir."

"In that case, Venerable One," concluded Menander, "you told me a lie when you said you were Nagasena. For I find no Nagasena. There is no Nagasena at all. Only a name."

Nagasena then asked Menander, "How did you get here, sir? By foot or in a vehicle?"

"In my chariot," replied Menander.

"If you came by chariot, sir, then show me this chariot. Is the axle the chariot?" asked Nagasena.

"Well, no."

"Are the wheels the chariot?"

"No."

"The body, the flag-staff, the yoke, the reins, the goad—are these the chariot?"

"No."

"Then is there a chariot apart from these things?"

"No, Venerable Nagasena."

"Then are not you, the King of India, also lying? For you said that you came by chariot, but are unable to produce anything. So what is this 'chariot'? It too is nothing but a name."

"No, Venerable Nagasena," conceded the King, "I am not telling a lie. For it is because of these parts that a chariot exists as a denotation, a conventional usage, a name."

"Exactly," replied Nagasena, "And so it is for me, sir. Because of my body, feelings, perceptions, mental impulses, and consciousness does Nagasena exist as a denotation, a conventional usage, a name. But ultimately there is no person at all to be found."

3

Truths

The Buddha died between two trees in a grove near Kusinara (present-day Kushinagar, India) at the age of eighty, according to the *Mahaparinibbana Sutta.*[1] His disciple Ananda and other followers were at his side. Although the Buddha provided forty-five years of teachings in-between his enlightenment and death, the awakened one offered one last opportunity for questions:

> It may be, *bhikkhus*, that one of you is in doubt or perplexity as to the Buddha, the *dhamma*, or the sangha, the path or the practice. Then question, *bhikkhus*! Do not be given to remorse later on with the thought: "The Master was with us face to face, yet face to face we failed to ask him." (Digha Nikaya [DN] 12.6.5)

The *bhikkus*, however, were silent. The Buddha offered two more opportunities for the followers to ask questions, though silence continued to be the only response. Thus, the Buddha shared his final words, illustrating his own susceptibility to the truth of *anitya*, impermanence, and imploring them to continue with conviction:

> Behold now, *bhikkhus*, I exhort you: All compounded things are subject to vanish. Strive with earnestness! (DN 12.6.8)

After the Buddha died, the First Buddhist Council was organized and convened by Mahakasyapa, one of the Buddha's chief disciples, and attended by Ananda and approximately five hundred other enlightened *arhants*. The gathering took place at the entrance of a cave in Rajaghra, three months after the Buddha's passing. The central purpose of the council was to preserve the *sutras*, teachings of the Buddha, and the *vinayas* (Sanskrit; also *vinayas* in Pali), the rules of conduct and discipline for monks and nuns. This meeting, said to have lasted seven months, established a tradition of oral transmission, from teacher to student, that continued in the many centuries following the Buddha's life and death. Some scholars doubt the historicity of such a grand, singular event where the voluminous teachings and disciplinary rules of the Buddha were recited and established all at once. The only historical sources describing this event derive from Buddhist canons. However, there is a high degree of concordance between Pali texts describing this event and Sanskrit traditions, leading other scholars to support the veracity of The First Buddhist Council. Regardless of the historical details, the century or so following the Buddha's death was marked by unity in the *sangha*, and a shared sense of responsibility for maintaining and transmitting the Buddha's teachings and practice guidelines.

A complex and often uncertain history of *sangha* splintering accompanied Buddhism's evolution as it spread and diversified across Asia and other parts of the planet over the last ~2,400 years. The first schism resulted from the Second Buddhist Council, an event that scholars agree took place in Vaishali (a city in present-day Bihar, India) during ~383 BCE. Shortly before his death, the Buddha shared that the *sangha* could dispatch with some of the "minor" rules for monastic discipline, but he never specifically delineated which such rules were dispensable. At the second council, a minority group of "elder" re-formist *bhikkus* charged that the *vinayas* practiced by other members of the *sangha* were insufficient, and that discipline needed to be tightened. However, this elder group's effort to modify and add rigor to the *vinayas* of the broader *sangha* were unsuccessful. The assembly of elder reformist monks separated and developed a set of *vinayas* with additional rules, forming a distinct sect called the Sthaviras. The broader community, which rejected the Sthaviras' stricter set of disciplines, came to be known as the Mahasamghikas.

In the subsequent centuries, the Sthavira and Mahasamghika schools each further fractured into subschools. It is commonly said that there were eighteen early schools of Buddhism in this eight-century period (from ~300 BCE to 500 CE).[2] The schools descended from the Sthaviras continued to maintain a rigid and expanded set of *vinayas* relative to the other schools and migrated southward on the Indian subcontinent and into Sri Lanka. The Sthavira lineage ultimately evolved into contemporary Theravada Buddhism, centered in Sri Lanka, which in turn came to prevail into other regions of southeastern Asia such as modern-day Thailand, Cambodia, and Myanmar. All the other early schools of Buddhism went extinct, though records for some schools remain intact in the Chinese canon. The Mahasamghikas are often implicated as the predecessors to modern Mahayana forms of Buddhism, though the development of this branch was more complicated as compared to the relatively linear evolution of the Sthavira into Theravada.

The Perfection of Wisdom

Theravada and Mahayana are commonly described as the two main branches of Buddhism in current existence.[3] Mahayana serves as the foundation, in terms of both philosophy and meditation, for the many contemporary forms of Zen and Tibetan Buddhism practiced today in many parts of Asia and the rest of the world. Some scholars suggest that Mahayana arose as a movement within the early Mahasamghika School sometime around the first century CE in India; others support a theory that Mahayana evolved as a pan-Buddhism movement with influences from many of the early eighteen schools. Early Mahayana followers practiced alongside other Buddhists in places such as monasteries and caves, but they followed different *sutras* and held distinctive philosophical views. In fact, most consider Mahayana to not be a separate and distinctive "school" of Buddhism, but rather a general set of ideas and principles, guided by particular *sutras*, that distinguish its adherents from those following other sets of teachings. Mahayana Buddhists often view their path as the "Greater Vehicle" to enlightenment, juxtaposing their approach against the Hinayana, "Lesser Vehicle," philosophies and practices of other Buddhists. Despite common misconception, modern Theravada is not generally considered as Hinayana among Mahayana adherents. Rather, Hinayana usually refers to the many early Buddhist schools with which Mahayana coevolved in its early development many centuries ago.[4]

The *Prajnaparamita Sutras* ("Perfection of Wisdom" *Sutras*) constituted central, defining teachings that guided the development of Mahayana Buddhism. These influential texts were composed on the Indian subcontinent between ~100 BCE and 600 CE. The *sutras* appeared in and were central to the Chinese and Tibetan canons. However, they are completely absent in the Pali Canon of Theravada Buddhism. Mahayana Buddhists claim that the stories of the Buddha presented in these *sutras* come from his later years in life, and offer teachings that are directed toward only the most high-level practitioners. Stories state that the *Prajnaparamita Sutras* were kept in monastic secrecy for centuries following the Buddha's death, waiting upon a period when the world was ready for these advanced teachings. In Buddhist legend, the *Prajnaparamita Sutras* were hidden and protected in a realm of *nagas*, mythological snake-like beings, for centuries. Nagarjuna, a highly influential second-century CE Buddhist monk from southern India, is said to have traveled to this mystical *naga* realm to retrieve the sacred *sutras* and return them to the world of humans.

The *Prajnaparamita Sutras*, and subsequent commentaries on those sacred texts (and commentaries on those commentaries, and so on), set forth philosophies and principles that would become distinctive, defining features of Mahayana Buddhism. Much of Mahayana centers on the *bodhisattva* (Sanskrit; *bodhisatta* in Pali) concept. In the early Buddhist schools and Theravada, the term *bodhisattva* was often used in a specific and limited fashion, referring to the pre-enlightenment phase of beings that would eventually become Buddhas, such as Siddhartha Gautama before his enlightenment under the Bodhi tree. In Mahayana, by contrast, a *bodhisattva* is anyone who has vowed to achieve enlightenment for the purpose of liberating all sentient beings in the universe from *duhkha* and *samsara*. The Mahayana *bodhisattva* is motivated by *bodhicitta* (Sanskrit), a spontaneous sense of universal and indiscriminate compassion for everyone, everywhere. For Mahayana Buddhists, it was insufficient (and in some ways selfish) for a person to aspire for and achieve individual liberation from *samsara* (i.e., become an *arhant*) for oneself while there continued to be countless other beings still suffering in *duhkha*. Thus, one who has taken the *bodhisattva* vow experiences *bodhicitta* and foregoes passage into *nirvana*, opting instead to continuously return through rebirths until all sentient beings in the universe are enlightened.

Specific *bodhisattvas* became the foci of devotional and worship practices as Mahayana spread across Asia, and served as symbols for specific qualities deemed desirable in Buddhist traditions. Avolikitesvara, the *bodhisattva* of compassion, first appeared in the *Saddharma Pundarika Sutra*, better known as the Lotus *Sutra*. This and other Mahayana *sutras* tell stories of Avolikitesvara's profound *bodhicitta* and tireless work to aid all sentient beings. Avolikitesvara and other *bodhisattvas* conduct their compassion work through *upaya* (Sanskrit), often translated as "skillful means," whereby they skillfully adapted their strategies to specific individuals and situations. The Lotus *Sutra* describes thirty-three different manifestations of Avolikitesvara, including both male and female versions of the *bodhisattva*, each identity specifically adapted to the minds of various beings that they helped. Avolikitesvara, usually male in most early Indian representations, evolved into the female Guanyin *bodhisattva* in China and other parts of eastern Asia. Avolikitesvara became Chenrezig in Tibet and is said to emanate through the Dalai Lama lineage. Jizo (known as Ksitigharba in Sanskrit) offers a second well-known *bodhisattva* example and also transitions between female, male, and gender-neutral forms in Buddhist iconography. Jizo is considered a *bodhisattva* guardian of youth and protector of sick and deceased children.[5] Small Jizo statues, often garbed in red cloaks, are common features of gardens and temples in modern-day Japan and other places where many Japanese people have migrated over the last century such as Hawai'i and the continental United States.

Mahayana prescribes a set of six *paramitas* (Sanskrit; *paramis* in Pali), usually translated as "perfections," for one pursuing the *bodhisattva* path to enlightenment.[6] The six perfections are presented in an ordered path but also are intended to mutually support one another. According to the *Astasahasrika Prajnaparamita Sutra*,[7] the earliest known Mahayana *sutra* (dating from the first century BCE), the six perfections include the following:

- *Dana Paramita*: perfect generosity, giving. The first perfection involves giving that derives from an authentic desire to benefit others, without expectation of reward or recognition. Nothing is expected in return. Charity work, done for the purpose of "feeling good about myself," is not *Dana Paramita*. Genuine generous acts offer portals to *bodhicitta* and experiential understanding of the *dharma*.

- *Sila Paramita*: perfect morality, discipline. In some senses, such as for renunciate Buddhist monks and nuns, *Sila Paramita* relates to following the specific disciplinary guidelines set out in the *vinayas*. More broadly, dedicated followers of the Mahayana path often take refuge in five precepts (no killing, stealing, lying, sexual misconduct, or intoxication) as a step toward establishing moral discipline and enlightenment. *Sila Paramita*, however, is not just about following prescribed sets of rules but also reflects the establishment of an ethical compass that results from Mahayana practice. An advanced *bodhisattva* is said to naturally respond correctly to all situations without having to consult a list of instructions.

- *Ksanti Paramita*: perfect patience, endurance. It is often said there are three dimensions to *Ksanti Paramita*. The first is the ability to endure personal hardship, continuing on the path to enlightenment despite one's ongoing *duhkha*. The second feature is patience with others, being tolerant and compassionate toward everyone that one encounters. The third dimension is patient acceptance of truth, as presented in the Four Noble Truths, and also the truth and self-realization of one's *duhkha* experience and the impermanent nature of all *dharmas*.

- *Virya Paramita*: perfect vigor, diligence. Development of a proper understanding of the Buddha's teachings leads one to advance on the path to enlightenment with confidence and passion. Inner, personal motivation is initially cultivated, followed by sharing that energy and motivation with others.
- *Dyana Paramita*: perfect concentration, meditation. The establishment of a clear, focused mind through meditative training is essential to all Buddhist paths. *Bodhisattvas* develop the ability to transfer the calm concentration and insight achieved in meditative practice to all other facets of their daily lives.
- *Prajna Paramita*: perfect wisdom. In Mahayana Buddhism, the attainment of perfect wisdom equates to a direct experiential understanding of the nature of reality and the universe, according to the Buddha's view expressed in the Four Noble Truths, Three Marks of Existence (*duhkha, anitya, anatman*), and the mutual cause-and-effect framework expressed in *pratityasamutpada*. In Mahayana, however, the concept of *sunyata* (Sanskrit; *sunnata* in Pali[8]), commonly translated as "emptiness," came to prominence as a central philosophical paradigm to express the perfection of the Buddha's wisdom.

The most popular *sutra* in Mahayana Buddhism is the *Prajnaparamitahrdaya* (Heart of the Perfection of Wisdom) *Sutra*, more simply and commonly known as the Heart *Sutra*. This short text shares a discourse between the *bodhisattva* Avolikitesvara and Sariputra, a follower of the Buddha who, though appearing as a wise *arahant* in Theravada *suttas*, is often depicted as a naïve question-asker in Mahayana *sutras*. The *sutra* offers a terse and somewhat cryptic explanation of *sunyata*, emptiness:[9]

> ... through the Buddha's inspiration, the venerable Sariputra spoke to holy Avolikitesavara, the bodhisattva, the great being, and said, "How should a noble son or noble daughter who wishes to engage in the practice of the profound perfection of wisdom train?"
>
> When this had been said, holy Avolikitesvara, the bodhisattva, the great being, spoke to venerable Sariputra and said, "Sariputra, any noble sons or daughters who wish to practice the profound perfection of wisdom should see this way: they should see insightfully, correctly, and repeatedly that even the five aggregates are empty of inherent nature."

Form is emptiness; emptiness is form. Emptiness is not other than form; form is not other than emptiness. Likewise, feelings, discrimination, compositional factors, and consciousness are empty. In this way, Sariputra, all things are empty; they are without defining characteristics; they are not born, they do not cease, they are not defiled, they are not undefiled. They have no increase; they have no decrease.

* * *

Nagarjuna further developed the *sunyata*, emptiness, philosophical concept that is central to the Heart *Sutra*. This second-century CE Indian philosopher is considered by many Mahayana followers to be second only to the Buddha in terms of contributions to Buddhist thought. In contrast to the Buddha's pragmatic outlook, Nagarjuna developed deep metaphysical theories that would come to captivate and influence the minds of his followers and much of the Mahayana Buddhist world. Like the Buddha, Nagarjuna engaged in philosophical debate with non-Buddhist philosophers and spiritualists of his time (e.g., those from the Nyaya school of Hinduism in second-century CE). However, Nagarjuna also debated and criticized the philosophies of other Buddhists (e.g., those from the second-century CE Sautrantika School that derived from the early Sthavira School) who held, according to Nagarjuna, imperfect views of the *dharma*. Nagarjuna developed the Madhyamaka, "Middle Way," branch of Mahayana philosophical thought, which underpins most modern-day varieties of Zen and Tibetan Buddhism. He wrote numerous dense tomes such as the *Mulamadhyamakakarika* ("Fundamental Wisdom of the Middle Way") that provided commentary and analysis of the philosophical concepts set forth in the *Prajnaparamitra Sutras*.

Sunyata, emptiness, is central to Madhyamaka philosophy. Nagarjuna's conception of *sunyata* transcends the *anatman* concept that is shared by all forms of Buddhism by claiming that all *dharmas* (i.e., everything in the universe, not just the "self" nature of human beings) lack any kind of intrinsic nature. According to this outlook, all things and phenomena lack any form of inherent essence, but rather have a completely conditional form of existence that is entirely contingent upon interactions with other *dharmas* (that are, themselves, also "empty" in this same way). The Middle Way philosophy's truth exists in-between the two extreme poles of eternalism (the view that *dharmas* have inherent essences) and nihilism (the view that *dharmas* have no form of existence whatsoever).

In the *Mulamadhyamakakarika* and other works, Nagarjuna framed the Middle Way philosophy as a dual thesis, commonly known as the "Two Truths." The first of these truths is *sunyata*, emptiness, which in the Two Truths framework is referred to as the ultimate truth (*paramartha satya*, Sanskrit). All *dharmas* lack an inherent, independent, existence; there is no fundamental essence in anything, anywhere, ever. At the same time, however, there is also the second truth of conventional truth (*samvriti satya*, Sanskrit). *Dharmas* do, in fact, have a conventional form of existence, but one that is entirely conditioned based on a variety of contextual factors. In his translation and commentary on the *Mulamadhyamakakarika*,[10] Jay Garfield, a contemporary Buddhist philosopher and Smith College professor, offered the following example:

> . . . when a Madhyamaka philosopher says of a table that it is empty, that assertion by itself is incomplete. It invites the question, Empty of what? And the answer is, Empty of inherent existence, of self-nature, or, in more Western terms, essence. Now, to say that the table is empty is hence simply to say that it lacks essence and importantly not to say that it is completely nonexistent. To say that it lacks essence, the Madhyamaka philosopher will explain, is to say, as the Tibetans like to put it, that it does not exist "from its own side"—that its existence as the object that it is—as a table—depends not on it, nor on any purely nonrelational characteristics, but depends on us as well. That is, if our culture had not evolved this manner of furniture, what appears to us to be an obviously unitary object might instead be correctly described as five objects: four quite useful sticks absurdly surmounted by a pointless slab of stick-wood waiting to be carved. Or we would have no reason to indicate this particular temporary arrangement of this matter as an object at all, as opposed to a brief intersection of the histories of some trees. It is also to say that the table depends for its existence on its parts, on its causes, on its materials, and so forth. Apart from these, there is no table.

For Nagarjuna and his followers, everything was subject to this "middle" form of existence at the nexus of ultimate and conventional truth, no exceptions. In the twenty-fourth chapter of the *Mulamadhyamakakarika*, Nagarjuna analyzed the precious Four Noble Truths of Buddhism and determined that they are also empty. He concluded that the achievement of Buddha-hood was empty, and that *nirvana* was empty. Nagarjuna went further to connect his analysis of *sunyata* back to the core Buddhist concept of *pratityasamutpada*, dependent arising, and concluded that even the emptiness concept itself was empty:

> Whatever is dependently co-arisen
> That is explained to be emptiness
> That, being a dependent designation
> Is itself the Middle Way
>
> (Mulamadhyamakakarika 24.18)

Understanding the complex interrelationships between the ultimate and conventional truth shaped many centuries of debate within Madhyamaka Buddhist circles. Gorampa, a fifteenth-century CE philosopher of the Sakya School of Tibetan Buddhism, argued that conventional truth was not a genuine truth—he charged conventional truth to offer nothing but illusion to ignorant minds, the source of the fogs of *samsara*. Tsongkhapa, a contemporary and philosophical opponent of Gorampa in fifteenth-century Tibet, countered that conventional truth was, in fact, a bona fide truth. Its validity, however, was dependent upon the wisdom of the thinker—it was a valid truth for those who see and understand the emptiness (i.e., ultimate truth) in all *dharmas*, but not for those who perceived themselves, objects, and so on as having some kind of inherent essence. Tsongkhapa's philosophical views and debate-methodological approaches came into prominence in Tibet,[11] and he founded the Gelug School of Tibetan Buddhism, the school of the Dalai Lamas. In verse 11 of *Praise of Dependent Arising*,[12] Tsongkhapa further legitimized conventional truth by relating *sunyata*, emptiness, to *pratityasamutpada*, dependent arising, and emphasizing the importance of the latter in the functioning actions of *dharmas*:

> For you, when one sees emptiness
> In terms of the meaning of dependent origination,
> Then being devoid of intrinsic existence and
> Possessing valid functions do not contradict.

Madhyamaka philosophers often deploy a mirage simile to describe the distinction between conventional and ultimate truth, and the centrality of perception in that process. The mirage of a water oasis in desert sands appears to be water, but is in fact empty of water—it is deceptive and, in that sense, false in nature. Yet a mirage is not absolutely nothing—it is in fact a mirage, in the form of a deceptive mental construct in the person's mind. But it is still not "real" water in any way. Contemporary Madhyamaka philosopher Jay Garfield modernized the analogy and expanded upon it to emphasize the validity of conventional truth:[13]

> Imagine three travelers along a hot desert highway. Alice is an experienced desert traveler; Bill is a neophyte; Charlie is wearing polarizing sunglasses. Bill points to a mirage up ahead and warns against a puddle on the road; Alice sees the mirage as a mirage and assures him that there is no danger; Charlie sees nothing at all and wonders what they are talking about. If the mirage were entirely false—if there were no truth about it at all—Charlie would be the most authoritative of the three (and Buddhas would know nothing of the real world). But that is wrong. Just as Bill is deceived into believing that there is water on the road, Charlie is incapable of seeing the mirage at all and so fails to know what Alice knows—that there is an actual mirage on the road, which appears to some to be water, but which is not. There is a truth about the mirage despite the fact that it is deceptive, and Alice is authoritative with respect to it precisely because she sees it as it is, not as it appears to the uninitiated.

Conventional truth and ultimate truth are both valid, according to Nagarjuna, the Dalai Lama, and other followers of this line of Mahayana Buddhism. However, referring to them as the "Two Truths" can be misleading, giving the false impression that they are unrelated, distinct truths. To clear up this confusion, a coin analogy is often deployed. Conventional truth is on one side of the coin, and it can be intellectually understood and experienced by anyone who has benefitted from the appropriate teachings. Ultimate truth (*sunyata*, emptiness) is on the other side of the coin, but it can only be truly understood through direct experience that comes with advanced meditative training. Both sides of the coin are there, but you can only ever see one side at a time. Conventional truth and ultimate truth are two conceptual identities for a single, valid truth.

The Truth in Aspens

Sometimes the truth hides. It hid from Siddhartha until his awakening experience under the Bodhi tree. It was always there, but the pre-enlightenment Buddha-to-be could not see it. Water in the air, condensing as fog, veils trees in a forest and obscures a walker's view. In the fog, a hiker cannot see that there is a majestic ancient tree—or, perhaps, a hungry bear—just around the corner. In these circumstances, the walker knows that their perception and perspective are limited, and the reasons why. Sometimes, however, hikers navigate paths without understanding all of the factors that limit their view. Fogs exist that remain invisible unless one goes looking for them, stumbles upon them, or are pointed out to them by teachers.

Aspens inhabit the Rocky Mountains. These iconic white-barked trees prevail throughout many parts of North America and offer deciduous contrasts to the coniferous firs, pines, and spruces that dominate landscapes in many parts of western North America. My in-laws inhabit the Rockies, in the state of Colorado; Stephanie, Amani, Hirut, and I visited them there for the winter holiday break during December 2012. Having recently received tenure, I was on sabbatical leave from OSU and taking some time for family and to advance my scholarly knowledge in Buddhism. In addition to visiting extended family, the sabbatical flexibility enabled me to care for Amani and Hirut in the afternoons after their half-days of morning kindergarten. In the evenings during the sabbatical, I would drive to Portland to learn about Tibetan Buddhist teachings at Maitripa College. During the 2012 holiday visit to Colorado, we were also joined by Steph's sister, Kristen, and her family. My in-laws, Mike and Jan Swenson, occupied a beautiful home in the depths of the Rocky Mountains that hosted us all. Mike picked our family up from the airport, and on the drive to their house we saw snow. We saw elk. We saw aspens.

Walks through the beautiful, snow-covered forest behind Mike and Jan's house were a standard feature of our annual holiday visits to Colorado. The morning after our arrival in 2012, after breakfast, the entire extended Swenson family bundled up and went outside to embark upon the hike. *Duhkha* was present. I had a headache, likely due to a combination of the unfamiliar and thin high-altitude air, the three beers from the night before, and Amani and Hirut's disrupted sleeping schedules that accompanied the time zone switch. They had been up much of the night. My tired and dehydrated brain produced a blend of discontent and irritability in my mind. All morning I kept to myself; I did my best to remain disengaged from the early-morning conversations as we started the walk down the mountain. All questions directed at me were answered with terse responses; I made no effort toward being a meaningful conversation partner with anyone. Any little thing would annoy me.

After a few frustrated looks from Steph, who inhabited her own *duhkha* caused in part by my disengaged behavior around her family, I bowed my head and took a deep breath, deciding to try my best to just get over it. I then looked up and ahead, seeing Amani and Hirut with their grandpa, Mike, a little further up the trail. While I had been wallowing in negativity, Steph's dad was enthusiastically engaged with our now five-year-old children. The heavy coat–bundled kids ran energetically from tree to tree, with bits of snow bouncing out of the accumulation in the curly black hair atop each of their heads. Mike did his best to keep up. The stocking caps that we insisted the children wear outside, and the source of protests and a near tantrum at the onset of the family walk, were now nowhere to be seen. *Grr.* I took another deep breath, resolved again to try to change my attitude, and then quickened my pace to join the trio up ahead while Steph turned her attention to her nearby mom and sister.

Mike smiled as I arrived, and he immediately engaged me in a conversation about trees. I noticed snow accumulating on his Chicago-style moustache as chipper words describing the different kinds of trees on their property emanated from his mouth. The semiretired business communications consultant had come to accumulate quite a bit of knowledge on the Colorado flora and fauna ever since he and Jan moved there from the Chicago suburbs a number of years back. Mike especially admired the aspens, and the conversation evolved into an aspen counting activity with Amani and Hirut as the group started its return journey to the house. The kids were challenged to count the number of the white-barked trees on the way back, with a promise of hot chocolate if they found more than ten. After a few isolated aspens contributed to the early tally, Hirut and Amani discovered a small copse of the thin trees which effectively ended the game and guaranteed warm cocoa back at the house.

After a few more days of hiking, watching basketball and football on TV, and trips out to restaurants, the visit wrapped up. On the morning of our return flight back to Oregon, Stephanie and I secured a cranky Amani and a crankier Hirut into their car seats and then Mike drove the four of us back to the airport for our super-early flight. Steph was in back of the SUV with the kids, and I was up front in the passenger seat drinking some much-needed coffee. A soothing Beatles song emanated from the vehicle's sound system, which combined with the warm internal temperature to lull the kids back to sleep. I was feeling sleepy, too. As we drove out of the subdivision, Mike returned to our previous conversation about trees. He told me about a fascinating article that he had read which explained that many aspen stands shared an underground root network beneath the soil, and that the constituent trees in the copse—appearing as multiple individuals when viewed from above ground—could all be considered one giant organism. Mike shared more details of the story, recounting how if one followed a root from the trunk of one tree stalk, it would lead to the trunk of another stalk in the same group. All of this was hidden to the human observer, however, because the root connection between trees only occurred, and was thus only visible, underground.

My mind was blown. Mike's words jolted me, catalyzing thoughts of *anatman* ("non-self") and *sunyata* ("emptiness") in my mind, concepts discussed in my Buddhism books and the Maitripa College classroom. After sharing the story, and a bit of back and forth with me about it, Mike's attention turned to the music, and he hummed along with the Beatles a little bit. My mind, however, roiled. I was initially embarrassed with myself, that as a biologist I had no idea about this amazing truth of aspens. I held an internal debate inside my mind about the possibility of building on the conversation with Mike, to extend it to the Buddha's teachings. After his humming, however, Mike transitioned to conversation with his daughter in the back about the Corvallis kindergarten experience for Hirut and Amani. The moment had passed. My mind, however, continued to dwell upon this unforeseen aspen insight that offered a marvelous biological metaphor for Buddhist ideas, taught to me by my father-in-law, Mike. After getting over the *duhkha* embarrassment held by the recent previous version of myself, and letting my mind work on this more, a sense of inspiration subsequently settled upon me as we arrived at the airport. Mike pulled his SUV to the curb at the drop-off zone and then popped out to grab our luggage, as Steph and I disconnected the half-asleep kids from their car seats and then the car seats from the vehicle. Everyone said their goodbyes to Mike, and I gave my father-in-law a bigger hug than usual.

4

Intersections I

Following the Buddha's principle of *pratityasamutpada*, it is perhaps folly to seek out a true "first" interaction between Buddhism and Western science. One might point to Greek philosophers, such as Pyrrho, the patriarch of skepticism, who accompanied Alexander the Great during his ~334 CE campaigns into central Asia and encountered a variety of gymnosophists ("naked wise men") that likely included followers of the Buddha's teachings. Some suggest that Buddhism had marked impacts on Pyrrho and his philosophy, noting his personal tendencies toward seclusion and equanimity, and the strong connections between basic features of Greek skeptical philosophy and the Buddha's three marks of existence (*duhkha, anitya, anatman*).[1] Parallels have also been made between Pyrrho's skepticism and the Two Truths doctrine.[2] Other scholars, however, are more dubious about Buddhism's impact on Pyrrho; they point to the early development of skeptic thought in Greece prior to the philosopher's Asian excursion and the translation difficulties expected to accompany interactions with the gymnosophists.[3] Beyond Pyrrho, there are other historical interactions between ancient Buddhists and Greeks, such as the Menander's Questions dialogues presented in Chapter 2, and the proselytizing efforts of the famous Buddhist King Ashoka of India who sent emissaries across different parts of Eurasia, including Greece, during the third century BCE. Such interactions between ancient Greeks and early Buddhists might have impacted the Western philosophical foundations of modern science, though more overt Buddhism and science interactions began much later, in the nineteenth century CE.

The "Buddhism and science" scene, though largely unfamiliar to most practicing scientists and broader swaths of contemporary Western society, has a large and expanding bibliography authored by Buddhist Studies scholars that are part of the liberal arts tradition in academia, along with some contributions from Buddhist monastics and occasional scientists. Much of the Buddhism and science dialogue has focused on neuroscience and physics, and revolved around Tibetan Buddhism. For example, *Buddhism & Science: Breaking New Ground* provided a pioneering analysis of Buddhism and science intersections that centered attention on the cognitive and physical sciences and featured writers from the world of Tibetan Buddhism such as Alan Wallace, Matthieu Ricard, Thupten Jinpa, and the Dalai Lama.[4] The nature of consciousness and the physical universe were also major (though not exclusive) shapers of the Dalai Lama's Mind and Life Conferences, featured in *The Universe in a Single Atom: The Convergence of Science and Spirituality*.[5] Although this thread of Buddhism and science dialogue is the most well-known and visible vector of this intersection, there exist many more and less well-known efforts that involve different varieties of Buddhism and different kinds of science. For example, Professor J. K. P. Ariyaratne from the Theravada Buddhist island of Sri Lanka began synthetic work on Buddhism and science in the 1980s, focusing on chemical processes and paradigms. In 2003, he published *Two Buddhis Sutras Viewed from Science*,[6] which evaluated chemical bonds, electrochemistry, and the scientific method itself from a Buddhist lens.

The intersection of Buddhism and biology has also been explored, with prominent examples arising again from the world of Tibetan Buddhism. Most notably, the Dalai Lama's *The Universe in a Single Atom: The Convergence of Science and Spirituality* offered a chapter on genetics and biological evolution that described dialogue from the 2002 Mind and Life Conference in Dharamshala, India. However, there exist many less well-known efforts aimed at synthesizing Buddhism and biology that incorporate perspectives from other Buddhist traditions. This chapter, and the following chapter, will focus on the manifold and often underexplored intersections between Buddhism and biology. A thorough tour through these previously explored crossroads is directly relevant and essential to the upcoming investigation of the Buddha's hypotheses in the second half of this book. This chapter will focus mostly on intersections with Theravada thinkers; the following chapter will focus mostly on Mahayana minds.

Colonial Connections

Sri Lanka, the teardrop-shaped island nation southeast of India, has been the epicenter of Theravada Buddhism for more than two millennia. During the third century BCE, King Ashoka of India sent his son, Arahantha Mahinda, as a Buddhist emissary to Sri Lanka. Shortly thereafter, Ashoka sent his daughter, Sangamitta, to the island; she delivered a sapling of the Bodhi tree,[7] which was subsequently planted in the then-capital city of Anuradhapura in north-central Sri Lanka. Sangamitta established the first order of Sri Lanka *bhikkunis*, Buddhist nuns, which flourished until the disappearance of the Theravada *bhikkuni* order in ~1000 CE, when the island was conquered by the Hindu Chola empire of southern India.[8] Nearly another thousand years would pass before the order of Buddhist nuns would be revitalized on the island. Sri Lanka and its Buddhist society suffered numerous invasions across the last two millennia, though none so great as that of the three waves of Christian colonialists from Europe; the island was under the rule of the Portuguese from 1505 to 1658 CE, the Dutch from 1658 to 1796, and the British from 1796 to 1948.

British colonizers imposed policies that forced Christian culture and religion upon nineteenth-century Sri Lankan society,[9] and also brought forth early European scholars of Buddhism. At this same time in history, the research and writings of Charles Darwin (such as the seminal *On the Origin of Species by Means of Natural Selection* in 1859[10]) catalyzed the evolution revolution that shook the world, and further strained the relationships between science and the Christian religions of Europe. Efforts to draw connections between Buddhism and Darwin's evolutionary theory appeared in the late 1800s, and involved both Buddhists and early European defenders of evolution. For some nineteenth-century Western intellectuals, Buddhism offered an exotic system of ethics that did not require the presence of a creator God and held features that seemed compatible with Darwin's then-new theory of evolution. Thomas W. Rhys David, professor of Pali at the University of London and the leading Theravada Buddhist scholar of his time, shared his outlook on the compatibility of Buddhism and evolution in an 1881 lecture:[11]

> And the more thorough-going the Evolutionist, the more clear his vision of the long perspective of history, the greater will be his appreciation of the strangeness of the fact that a theory [Buddhism] so far consistent with what he holds to be true should have been possible at all in so remote a past.

Edwin Arnold, a nineteenth-century British poet and journalist, found connections between Buddhism and Darwin's evolutionary science. This scholar traveled extensively around Sri Lanka, India, and other parts of Asia. Arnold published *The Light of Asia* in 1871, an account of the life and teachings of the Buddha prepared for Western readers. This bestselling book captured the fascination of European and American societies in the late 1800s, bringing widespread Western attention to the Buddha's story. Arnold visited Bodhgaya, India, in 1886 and found the place of the Buddha's enlightenment in a sad state of deterioration, and under the control of a Hindu priest; he reported on this situation in the *Daily Telegraph* of London and sparked interest in restoration efforts. Arnold connected Buddhism and evolution in an 1889 lecture during a visit to the Imperial University of Tokyo:[12]

> ... when Darwin shows us life passing onward and upward through a series of constantly improving forms toward the Better and the Best, each individual starting in new existence with the records of bygone good and evil stamped deep and ineffaceably from the old ones, what is this again but the Buddhist doctrine of *Dharma* and *Karma*?

The British biologist Thomas Huxley, nicknamed "Darwin's Bulldog," was professor of natural history at the Royal School of Mines and a staunch advocate for Darwin. He was a critic of Christianity but found appealing features in Buddhism. Though slow to accept Darwin's gradualism and the entire idea of natural selection, Huxley proactively promoted biological evolution through a famous 1860 debate with Samuel Wilberforce, a bishop in the Church of England and opponent of Darwin's then-new theory. Huxley worked to integrate Darwin's theory into the curriculum of the British educational system. He coined the term "agnosticism" in 1869 and clarified its function as a method of science rather than a creed. Huxley wrote,[13] "In matters of the intellect, do not pretend that conclusions are certain which are not demonstrated or demonstrable." Thomas Huxley, the champion of Darwin and father of agnosticism, wrote favorably about Buddhism in an 1894 essay:[14]

> A system [Buddhism] which knows no God in the western sense; which denies a soul to man; which counts the belief in immortality a blunder and the hope of it a sin; which refuses any efficacy to prayer and sacrifice; which bids men to look at nothing but their own efforts for salvation; which, in its original purity, knew nothing of vows of obedience, abhorred intolerance, and never sought the aid of the secular arm; yet spread over a considerable moiety of the Old World with marvelous rapidity, and is still, with whatever base admixture of foreign superstitions, the dominant creed of a large fraction of mankind.

These three late nineteenth-century British-born scholars, representing different academic disciplines and mindsets, each found potential promise at the Buddhism-science interface. It is clear from their quotations, however, that their understandings of the Buddha's teachings and Darwin's theory were often incomplete and/or inaccurate. For example, when Edwin Arnold spoke of Darwinian evolution in terms of "constantly improving forms toward the Better and the Best," the contemporary reader versed in modern evolutionary science will quickly find fault with the notion that there is some universal "Best" ideal form (e.g., the human form) toward which evolution marches. A Buddhist studies scholar of today likely has doubts about how well Thomas Huxley understood the Buddha's teachings in their "original purity." These deficiencies, however, are not a surprise; both Buddhism and Darwinism were new concepts that late nineteenth-century British and other Western societies were still struggling to digest and comprehend.

Stories from Sri Lanka: Anagarika Dharmapala

In 1864, in the southern seaport city of Matara in Sri Lanka (then known as Ceylon under British rule), Don David Hewavitarne was born into a middle-upper-class Sinhalese family that spoke English but was also Buddhist.[15] Like other merchant-class children growing up in nineteenth-century Sri Lanka, Don David was schooled in the Christian education system established by the colonial administration in power. Later in life, Don David would change his name to Anagarika Dharmapala and become an international champion of Buddhism and activist against British imperialism in Sri Lanka. He would also write on the parallels between Buddhism and biological evolution.

Don David loved reading the Bible. In his diaries,[16] he commented on the beauty of the rhythmic diction and poetry in the book's written word. However, Don David also read the book with a critical eye; he often fell under the physical abuse of headmasters when he challenged ideas found in the Bible or made comparisons to Buddhism. The young man made a habit of skipping his classes at St. Thomas's College during the holiday of Vesak, a celebration of the life, enlightenment, and death of the Buddha, and instead visited the Kotahena Buddhist temple near his parents' home in the suburbs of Colombo. The days of playing hooky resulted in canings from the St. Thomas's headmaster but also brought Don David closer to his ancestral Buddhist roots.

At the Kotahena temple, Don David was introduced to Migettuwatte Gunananda, a renowned Theravada monk who openly and bravely critiqued and criticized the Christian beliefs of Sri Lanka's colonial overseers. Gunananda previously founded the Society for the Propagation of Buddhism in 1862, and he established a printing press that produced pamphlets challenging the tenets of Christianity. Don David routinely attended Gunananda's public talks at Kotahena on his way home after school and increasingly became inspired to follow and promote the teachings of the Buddha.

As the crowds gathered and Gunananda's public talks increased in size, the British colonial authorities took notice. A public event was organized in 1873, and Gunananda and his fellow Buddhist agitators were challenged to participate in an open debate—to be held in Panadura, Sri Lanka—with Christian theologians. The Christian debate opponent of Gunananda was Reverend David da Silva, a Sinhalese-born convert to Methodism. In the debate, da Silva condemned the core Buddhist doctrine of *anatman*, "nonself"; according to Buddhism, da Silva declared, humans lacked an immortal soul and were thus "on a par with the frog, pig, or any other member of the 'brute' creation."[17] da Silva argued that the rejection of a soul and equation of humans to "lower" life forms made Buddhism a religion without a code of morality, and one favored by villains. When Gunananda's debate turn came, the monk criticized the Christian priest's mispronunciation and misunderstanding of Buddhist *suttas*, suggesting that da Silva, the adopter of the colonialists' foreign religion, did not understand the Buddha's teachings on the nature of personhood (i.e., *anatman*). Gunananda proceeded to enumerate contradictions found in the Bible, such as the story in Exodus where God instructed the Hebrews to mark their doors with blood so that they would be passed over when the first-born children of Egyptians were killed. If God were omniscient, Guananda argued, why would this be necessary? The three-day debate drew crowds numbering in the thousands, with many onlookers reporting that Gunananda had clearly won the debate. The Christian opponents, however, also claimed victory. The Panadura debate became an inspirational pivot point key to the revitalization of Buddhism in British-colonial Sri Lanka.

News of the controversial "Great Debate" in Sri Lanka rippled across the oceans and made its way into European and American newspapers. The story caught the attention of Colonel Henry Olcott, a New York veteran of the American Civil War who was one of the founding members of the Theosophical Society, an esoteric group formed in 1875 that sought connections between philosophy, science, and spirituality. The society described itself as "an unsectarian body of seekers after Truth, who endeavor to promote Brotherhood and strive to serve humanity."[18] A Russian occultist named Madame Helena Blavataski cofounded the Theosophical Society with Colonel Olcott; this pair traveled to Sri Lanka in 1880 to support the island's upstart Buddhist revivalism movement. Both Olcott and Blavatsky officially converted to Buddhism in a refuge ceremony shortly after arriving on the island, an action that endeared the two Westerners to the people of Sri Lanka, including Don David Hewavitarne.

Don David developed relationships with the Colonel and Madame during their first visit to Sri Lanka in 1880, and those relationships would further bloom in the years ahead. The two Theosophists returned to the island in 1884, and Don David joined them on their subsequent voyage to southern India; there the young Sri Lankan man was officially inducted into the Theosophical Society. In India, Olcott and Blavatsky encouraged Don David to study Pali and Buddhist philosophy, motivating him to return to his homeland and improve his knowledge and understanding of his native Buddhism and promote the Buddha's teachings across the Sri Lankan countryside. During this time, Don David Hewavitarne changed his name to Anagarika Dharmapala ("homeless truthseeker").[19] This label change, however, went beyond mere name; his new "Anagarika" path essentially created a new "middle path" social category of people who studied Buddhist texts and meditated like a monk but also stayed socially active in the world, unlike most monks. Dharmapala started to wear robes, though they were white instead of the reddish-yellow robes commonly worn by Theravada monastics. Dharmapala traveled by carriage around the Sri Lanka countryside, joined by Olcott and his Theosophist colleague C. W. Leadbetter, to learn more about the plight of the island's people under Christian colonial rule and to revive Buddhism across the island.

In 1889, Dharmapala joined Colonel Olcott on a lecture tour in Japan and initiated dialogue with Mahayana Buddhists to advance his promotional efforts of Buddhism to international stages. This experience inspired the Anagarika on a new mission to unite Buddhist traditions around the world. Dharmapala attended the World's Parliament of Religions in Chicago during 1893, and there attracted substantial attention as a captivating communicator of Buddhist teachings for Western ears and eyes. A letter published in the *St. Louis Observer* on September 21, 1893,[20] described Dharmapala's appearance and lecture style:

With black curly locks thrown from the broad brow, his clean, clear eye fixed upon the audience, his long brown fingers emphasizing the utterances in his vibrant voice, he looked the very image of a propagandist, and one trembled to know that such a figure stood at the head of the movement to consolidate all the disciples of Buddha and spread "the light of Asia" throughout the civilized world.

Dharmapala's speeches centered on three pillars that together supported a central mission to restore Buddhist identity and pride. The Anagarika first argued that, unlike Christianity, Buddhism was a spiritual tradition entirely compatible with modern science; he specifically pointed out the parallels between the Buddha's teachings and Darwin's theory of evolution. Dharmapala's second pillar rested on the compassion and love of the Buddha's ethical teachings, contrasting them with the ongoing suffering experienced by the Sri Lankan people at the hands of Christian colonizers. Third, the Anagarika emphasized that Buddhism was a religion of optimism and activism, not one of negativism and pessimism (a common charge levied by Christian missionaries).

In addition to a being a skilled orator, Dharmapala was a prolific essayist.[21] He wrote on fundamental Buddhist doctrines,[22] the associations between Buddhism and Christianity, and the intersections of Buddhism and science. On the specific topic of biological evolution, Dharmapala organized his thoughts on the relationships between the teachings of the Buddha and the teachings of Darwin in a 1924 essay titled "Evolution from the Standpoint of Buddhism."[23] Dharmapala wrote:

> The teachings of the Buddha are very little known in the west, and now that science is making great strides it is proper that Americans should know of the attitude that the Buddhist takes regarding the Darwinian exposition on the origin of species. Buddhists are taught not to believe dogmas and unscientific beliefs, and that belief that does not rest on the basis of the immutable law of cause and effect is rejected as unscientific.

In the essay, Dharmapala identified parallels in the then-recent debates between science and Christianity with centuries-old philosophical disagreements between Buddhist and Brahmanical views, involving questions relating to the nature of the universe. He pointed to the *Agganna Sutta*,[24] which described the Buddha's teachings on the nature of the universe, espousing a view whereby the cosmos constitutes a continuously changing and unfolding process involving infinite cycles of expansion and retraction, as opposed to "the Brahmanical view which posited a creator in the person of the Creator Brahma." Dharmapala extended this concept of change to the human and atomic levels, and connected it to *anatman*:

> The atom is changing, the universe is changing, the human body is changing with lightning rapidity, consequently there is nothing to call mine in the cosmic plane.

An echoing theme in the Buddhism and science dialogue, usually arising from the side of Buddhists describing their outlooks on science, is that the Buddha knew and understood realities of the universe many centuries prior to science's lagging efforts. The Buddhist view also often highlights perceived incompleteness in scientific characterizations, often pointing out features relating to human rebirth that continue to be largely discounted by most Western scientific minds, yet also believed by most Buddhist minds today. In "Evolution from the Standpoint of Buddhism," Dharmapala further wrote:

> Biology became a recognized science in the west only in the second or third decade of the nineteenth century, while the Buddha taught the biological view of life 2,500 years ago showing how the consciousness associates itself with the germ cell in the mother's womb, and how evolution takes place week after week of the dying man ceases only to be reborn by the force of karma in another life in the germ cell brought into existence by sexual contact of the parents. The rebirth of consciousness is explained in the Abhidharma books,[25] which are still sealed to western scholars.

In this essay and other writings,[26] Anagarika Dharmapala remained a steadfast champion of the Buddha's teachings, making positive comparisons between Buddhism and scientific discoveries as a weapon against the Christian beliefs of the colonial oppressors of his homeland. At the same time, Dharmapala also highlighted his views on the limitations of science, which he saw as a fledgling tradition of logic and reason as compared to the 2,500-year legacy of Buddhism. There are many features of these early twentieth-century essays with which modern scientists will certainly find fault, such as the notion that human consciousness somehow associates with germ cells,[27] for which there is no scientific evidence. However, other Buddhism-science connections made by Dharmapala deserve closer attention. For example, the Anagarika linked the Buddhist principle of *pratityasamutpada* (dependent arising), which negates any true first causes or origins, with the idea that there was no true "first origin" of humans on earth mediated by a Creator God, which was a central point of controversy in the Evolution-Creation debates of the late nineteenth and early twentieth centuries.[28] Dharmapala also reinforced the Buddhist ideal of believing things based on one's own experiences, logic, and reason rather than following dogmatic belief systems.

The Sri Lankan man who changed his name from Don David to Anagarika Dharmapala was an early thinker on the intersection of Buddhism and biology, at a time when Darwin's theory was new and still gaining acceptance and when Buddhism was just becoming known to the Western world. Dharmapala had a massive impact on the revitalization of Buddhism in Sri Lanka and its spread across the world. He became a bane of the British colonial powers in Sri Lanka and a leader in the reclamation and restoration of Buddhism's historical locations in India, such as the site of the Buddha's enlightenment in Bodhgaya, through his founding of the Maha Bodhi Society in 1891, which still operates today.[29] In 1933, Anagarika Dharmapala became a fully ordained *bhikku*, changed his white robes for orange robes, and then passed away later that same year.

More Stories from Sri Lanka: Bhikkuni Kusuma

Four years before Dharmapala's passing, a baby girl was born in the suburbs of Colombo, Sri Lanka. Though you won't usually see her name featured in any of the prominent "Buddhism and science" literature, she built and led the first biological laboratory at one of Sri Lanka's premier colleges, and she was awarded prestigious scholarships to pursue graduate studies in molecular biosciences in the United States. She resurrected the order of *bhikkunis*, Buddhist nuns, after a millennium-long absence in the conservative Theravada nation of Sri Lanka. This girl, named Kusuma Guanawardene, would come to be known as Bhikkuni Kusuma: a humble nun and pioneer of Buddhist feminism.

In 1939, while Kusuma was a ten-year-old girl, World War II erupted and she moved away from the suburbs of Colombo;[30] her family was concerned about the potential perils of Japanese bombs targeting the island nation's coastal capitol. Her father owned a one-hundred-acre rice paddy estate in the inland Kurunegala District, which is where Kusuma and her family lived for a year. There, the young girl played with her siblings in the paddy fields and sneaked into the nearby jungle to climb the famed Weeraba Rock (a.k.a. the Sigiriya Lion Rock) despite the warnings of her parents about serpents and other wild animals. Kusuma captured a baby rabbit during one of her visits to the jungle and brought it home to rear in a makeshift cage composed of a wooden box and wire mesh. One day, Kusuma discovered that her pet had died; she found it in the cage with wide-open, staring eyes. Her shrieks attracted neighbors; they informed her that a serpent likely killed the rabbit, and that it might still be lurking nearby. Shortly thereafter, Kusuma saw a cobra slithering around the cage, and she froze with fear. One of the neighbors grabbed the deceased rabbit out of the cage and threw it toward the snake. Everyone watched with fascinated horror as the cobra consumed the small mammal over the next half hour. After the rabbit was fully engulfed by the serpent, one of the neighbors grabbed a pole and beat the now engorged and slow-moving cobra to death. This experience of *duhkha*, which Kusuma described as a "terrible cruelty,"[31] stunned the young *bhikkuni*-to-be with grief and fear, and sparked in her a fascination with the nature of life and the nature of suffering.

When the war ended, Kusuma and her family returned to the Colombo suburbs. She enrolled in Ananda College, which at the time was the leading boy's school in Sri Lanka. Kusuma was one of only a few girls in a class of around forty-five total students and suffered constant jeers, whistles, and paper balls thrown at her by her male classmates. She studied at Ananda College for eight years, pursuing a demanding academic path toward medicine. Kusuma passed the theoretical and practical written medical examinations required to advance on to medical school, but she failed to pass the oral exam. Her professors at Ananda College, all men, subsequently encouraged her to proceed in the study of languages (Sinhalese, Pali) and "leave the study of Medicine to the boys of Ananda College."[32]

Kusuma rejected this patronizing advice; instead, she gained the necessary credentials to teach at the college level and secured a faculty position at Ananda College. There, she built from scratch the college's first Science Laboratory. She obtained all of the necessary equipment, chemicals, and posters necessary to conduct biological research at the college. Her lab was decorated with pictures of prehistoric animals, the anatomy and physiology of human organs, and Darwin's theory of evolution. In addition to offering the college's first practical lab environment for experiments to be conducted at Ananda College, Kusuma also organized the inaugural Science Exhibition, which attracted hundreds of adults and children from across Sri Lanka to see and experience science in action.

Kusuma's pioneering science activities in Sri Lanka helped her land her two scholarships, one from the Asia Foundation and a second from the National Science Foundation of the United States, to pursue graduate studies in America. In 1969, she joined the Molecular Biology graduate program at Ball State University in Muncie, Indiana, a decade before the Dalai Lama's brother would establish the Tibetan Cultural Center in Bloomington. The research equipment and facilities at Ball State amazed Kusuma; she learned how to use and then applied the cutting-edge lens of phase-contrast microscopy in her research. Midway through her graduate studies in Muncie, however, Kusuma's inspiration began to fade. She increasingly doubted science's ability to answer the big existential questions that occupied her mind; she asked her American mentors:

"What is the beginning of life?," "Why are we born?," "Why do we all have to die?," etc. But when I realized that our daily regimen of studies was not giving the answers to any of these questions, I began to question our Professors on these lines since they were the "experts" in studying the molecular basis of life. However, the only reply was, "Science has yet to discover all that!"[33]

The disillusioned young woman decided to turn elsewhere to seek the answers to her questions. Kusuma happily left her American science graduate program to return to Sri Lanka and start a new academic path focused on Buddhist philosophy and practice. At the age of forty-four, she obtained a bachelor's degree in Buddhist philosophy and English at the Sri Jayawardenepura University, followed by a master's degree thesis focused on *Satipatthana*, the foundation of mindfulness, and *Vipassana* meditation. She was awarded a grant from Taiwan to publish and freely distribute her master's thesis. In returning to her native Buddhism, Kusuma discovered and maintained the sense of satisfaction and deeper meaning that was lacking in her previous scientific studies:

Thus I finally found the answers to the questions that were bothering me for a long time. I found that understanding of the concept of *anatta* (Nonself), made me feel lighter and more kind and friendly towards the world.

Life and living definitely became much less of a challenge because I did not react too harshly to the problems of daily living. I also became more and more motivated to study Buddhist Philosophy and with my practice of Meditation, I soon learnt the art of integrating *Dhamma* with day to day living.

Kusuma continued her study of Buddhism at the doctoral level, turning toward an investigation of the history, experiences, and status of *bhikkunis*, Buddhist nuns, in India and Sri Lanka. In her final oral examination for the PhD, it was clear that Kusuma knew and understood more about the matter of Vinaya (monastic rules) for *bhikkunis* than her ordained *bhikku* examiners; she was awarded the doctoral degree and launched into a path that would ultimately result in the restoration of the *bhikkuni* order in Sri Lanka.

After achieving the PhD, Kusuma continued on as a teacher at Sri Jayawardenepura University and became increasingly concerned with the status of female followers of the Buddha. Some women in Sri Lanka, including Kusuma's own sister, became members of the *Dasasil Matas*, "ten-precept holding" nuns, who were not officially recognized as *bhikkunis*; the official lineage of Buddhist nuns disappeared on the island in ~1000 CE. Kusuma learned more and more about the struggles of these unrecognized nuns who usually lived in remote regions of the island, and she received little to no support from the government. Over the course of a decade, Kusuma interacted with numerous *Dasasil Matas*, ordained Theravada *bhikkus*, and other people in Sri Lanka who increasingly supported her work. This research evolved into a cause focused on returning the ordained *bhikkuni* order to Sri Lanka. She encountered substantial resistance from conservative Theravada *bhikkus* who claimed that the *bhikkuni* lineage was forever lost and considered Kusuma to be an activist and an agitator. But Kusuma persisted. She trained further with the *Dasasil Matas* in Sri Lanka and eventually joined their order. She spent time in the Kwanik-Ku temple in Seoul, Korea, where she learned from the Mahayana *bhiksunis* of the Chogya Order, discovering that the *vinayas* and ordination procedures for Theravada and Mahayana nuns were essentially identical. Her further research revealed that Mahayana and Theravada nuns could both trace their lineages back to Bhikkhuni Maha Prajapati, the first woman to be ordained by the Buddha. Kusuma made plans to go to Sarnath, India, the place of the Buddha's first teaching, to receive official ordainment as a *bhikkuni*.[34]

The Maha Bodhi Society, founded decades earlier by Anagarika Dharmapala,[35] supported Kusuma's mission to restore the *bhikkuni* lineage to Sri Lanka and offered to host the ceremony at Sarnath in India, where the *bhikkuni*-to-be stayed at the Anagarika's former residence. The ordination was received from Mahayana Chogya Order *bhiksunis* and *bhiksus* who traveled to India from Korea for the ceremony. On December 8, 1996, the official ordainment ceremony took place at the location where the Buddha delivered his first teaching. There were large crowds carrying Buddhist flags, Sri Lankan pilgrims, and on-looking *bhikkus* from Sri Lanka, Thailand, and other nations to witness the historic event. Bhikkuni Kusuma described her ordainment in 2008:[36]

> There were the ordaining masters, both *Bhikkus* and *Bhikkunis*, of the Korean *Sangha* and an interpreter to translate from Korean to the English language. It was my sacred duty to translate the ongoing procedure from English to Sinhala for the other nuns seeking ordination. The ceremony took eight hours and the discipline and formality was austere. My knees ached by bleeding and kneeling. I can see the scars even now as a reminder of this most auspicious day.

Kusuma Guanawardene, a woman who found answers to her scientific questions in the teachings of the Buddha, became the first ordained Theravada *bhikkuni* in nearly one thousand years. Today, her status as an official *bhikkuni* remains controversial and unrecognized by many *bhikkus* in Theravada nations such as Sri Lanka, Thailand, and Myanmar. Some claim that the Mahayana ordainment received from the Korean *bhiksunis* was illegitimate, despite Bhikkuni Kusuma's argument that all ordained Theravada and Mahayana Buddhist nuns can trace their *dharma* lineage to Maha Prajapati, the maternal aunt of the Buddha.

After the international headlines and fame that accompanied her historic ordainment, Bhikkuni Kusuma continued to support the *bhikkuni* cause in Sri Lanka but also sought a simpler life where she could offer practical help to people. She invested fifteen years in building the Ayya Khema Meditation Centre in Horona, Sri Lanka; the center was named after a German-born member of the *Dasasil Matas* with whom Bhikkuni Kusuma developed a strong friendship and shared a passion for advancing *bhikkuni* rights. This historic reviver of the *bhikkuni* order in Sri Lanka quietly continued to offer meditation instruction and guidance to local and international guests of her modest meditation center in the southern suburbs of Colombo until her death in 2021, at the age of 92.

Color-Blind

In 2011, Oregon State University awarded me tenure. During the preceding five years, I had developed a high-powered and well-funded research program that continued to make advances in understanding spontaneous mutation processes in our tiny worms, now leveraging new cutting-edge DNA sequencing technologies capable of detecting DNA changes across entire genomes. This was only possible, however, through the formation of a stellar research team that included a biotechnology expert (Dana Howe) with unparalleled molecular lab skills, a computer programmer (Larry Wilhelm) able to create new analytical workflows capable of handling the oceans of DNA sequence data, a mathematician (Peter Dolan) who thought nonstop about the nature of error and challenged us to look at the data with new critical lenses, and a driven undergraduate (Samantha Lewis) who performed essential verification experiments for this new and untested kind of DNA sequence data and would go on to become a junior professor in the Department of Molecular and Cellular Biology at the University of California, Berkeley. This research, culminating in a high-profile 2009 research article in *Proceedings of the National Academy of Sciences USA*,[37] also required collaborations with colleagues at Indiana University, University of Florida, and University of New Hampshire. I couldn't have done it by myself.

The tenure achievement sparked a new bravery in me to incorporate the study of Buddhism into my professional world. However, I needed teachers. I had learned much by reading books and delving deeper into the world of Buddhist philosophy over the last eight years, but the knowledge gained was limited. I had no one to talk to about it. I had done research and discovered a number of nearby Buddhist temples and meditation centers around Oregon, but I didn't know quite where to start. Despite the tenure-induced bravery, my science ego was still afraid of appearing as a novice in the world of Buddhism, which I most certainly was.

I decided to start local. A quick online search of the Oregon State University website led me to James ('Jim') Blumenthal. He was an associate professor of Buddhist Studies, and his website revealed him to be a person with deep experience in the world of Tibetan Buddhism. His online university profile picture revealed a curly-haired white guy in a green button-down shirt. Looking back, I remember thinking to myself that the color of his skin didn't matter; a Buddhist scholar was a Buddhist scholar. As I write this one decade later in 2021, I now know that it did. Jim offered the possibility of entering the world of Buddhism through a comfortably white, academic avenue. I was still very nervous about reaching out to him, and imaginary critical voices of evolutionary biologists continued to plague my mind despite my new tenured status.

I spent three days drafting email messages to Jim, revising and re-revising. Some drafts were short; others were way too long. I was obsessed with getting it just right, explaining to him my thoughts and interests at the Buddhism and biology intersection, but also not wanting to come across as pompous or a know-it-all scientist. One morning, after a sleepless night spent agonizing over the email message and whether or not to send it, I clicked the "send" button on the latest draft and then took a deep breath. Less than five minutes later, Jim had replied and suggested that we meet at Tarn Tip, a local Thai restaurant, for lunch that day. Anxiety and fear washed over me. I stood up from my desk chair and started to pace around my office. There was no time to prepare. I took a deeper breath and sent a quick reply to say thanks, and that I would meet him at the restaurant at 11:30.

Upon arriving to Tarn Tip, I could see Jim through the window; he was already seated and staring into a cup of tea. I gestured to the restaurant host that I was joining Jim and walked to the table to introduce myself. Jim looked up from the steaming cup and smiled at me. He was now balder than the version of himself shown on his online faculty profile; he wore a plain navy-blue button-down shirt with a collar. Nothing suggested "Buddhist" in his appearance or attire. We quickly launched into a conversation that popped around from Buddhism to biology to the nature of fatherhood to the notion of "birth." He challenged me. Jim's philosophy-rooted mind lobbed unfamiliar words and phrases at me throughout our dialogue, such as "soteriology" and "syllogistic strategies." One conversation thread that began focused on the shared value of direct experience in Buddhism and science careened toward Jim questioning why I didn't believe in rebirth when there were people out there who claimed to have directly experienced previous lives in near-death experiences. I reflexively responded with typical science skepticism regarding the possibility that these "direct experiences" might in fact be mere delusions and that they were in fact unverifiable due to the subjective nature of the evidence. Jim listened and smiled at me through the entire conversation.

As we finished our lunches, I queried Jim about the best way to gain deeper and more direct understanding of Buddhist teachings. His initial reply was simple: "Go to Portland." Jim shared that he was moonlighting at a small, up-start Buddhist college in Portland that followed the Gelug school of Tibetan Buddhism, the school of the Dalai Lamas. He elaborated that one of his good friends and a brilliant monk originally from Nepal, Yangsi Rinpoche, was the president and spiritual leader of this small center for Buddhist academics, Maitripa College. Jim encouraged me to take a sabbatical in the Maitripa College environment during the academic year ahead. I decided, then and there, to do just that.

* * *

On my first visit to Maitripa College to work out the details of my sabbatical, Yangsi Rinpoche was not there to meet me. This was a surprise. I was also expecting Jim to be there; he was nowhere to be found. Instead, I was greeted by a twenty-something academic advisor, garbed in a business-casual skirt and sweater. She was pleasant and welcoming, but not the sort of person with whom I was expecting to interact. I communicated my sabbatical vision to the young advisor, sharing that I wanted to sit in on upper-level Buddhist philosophy classes, do research in their library, and have opportunities to meet with Yangsi Rinpoche here and there to discuss the intersections of Buddhism and science. The young advisor, however, informed me in a polite but direct tone that I would need to take beginning-level Buddhist philosophy and meditation classes before moving to the advanced Buddhist philosophy. She also informed me that I would need to pay tuition for those classes, and that Rinpoche could not commit to any meetings because he had a very busy schedule.

My ego fumed with feelings of disrespect. My mind raced. None of this was going the way I had envisioned, and I started to wonder if all of this was a mistake. Thankfully, just as I was about to rudely storm out of the meeting, Jim arrived. He was wearing the exact same navy-blue shirt that he wore that first time we met at Tarn Tip. He read the situation quickly and calmed me down with his soft smile and a joke about his chronic tardiness. He explained to the advisor that I already had some basic Buddhism knowledge and suggested a middle path whereby I would simultaneously take the introductory and advanced courses. The advisor replied, "This is not how Rinpoche thinks we should do things." Jim gestured to her with a playful apologetic shrug and a smile. The academic advisor flashed us both a look of resigned resignation with a raised eyebrow and enrolled me in the two philosophy courses. We agreed upon a title of "Visiting Research Professor" to accompany my sabbatical, and that I would pay 50 percent tuition for the classes I sat in on during the sabbatical. It was well worth it.

During the 2012 fall semester at Maitripa College, I attended Introductory Buddhist Philosophy class from 4:00 to 6:50 and then the Advanced Buddhist Philosophy class from 7:00 to 9:50. The evening classes in Portland allowed me to spend time with Amani and Hirut, then kindergartners, during the morning hours and get them to and from school. Although Maitripa College gave me the title "Visiting Research Professor" during the sabbatical, Yangsi Rinpoche treated me as a regular student. I attended all of the monk's classes and did all of the assigned work. Around twenty classmates joined me in the introductory philosophy class; eleven or twelve were in the advanced class. The student population was generationally diverse. They included elder renunciates with shaved heads and garbed in Buddhist robes, middle-aged parents wearing hoodies and other typical Western clothes (like me), and young-adult hipsters with tattooed arms and colorfully dyed hair. Although there were a few Asian students, most were white.

Class periods began when Yangsi Rinpoche walked into the room; the students would then stand and bow with prayer hands in the direction of the maroon-robed teacher, a toweringly tall forty-something brown-skinned man with sharp, angular facial features. As Rinpoche walked in the room, the students' prayer hands would follow his path and remain pointed toward him until he sat down at the head of the classroom. Jim was usually there, too, at the head of the class alongside Rinpoche's seat. Jim bowed with the students. I did this, too, but always felt very weird about it. The students returned to their seats only after the tall, towering monk sat down at the front of the classroom. He always took a moment to appreciate the vase of flowers and cup of tea that the students always had ready for him. Dedicatory verses and chanting followed, prior to the commencement of Rinpoche's lesson. Most class periods involved the monk and/or Jim Blumental lecturing on assigned readings, focusing on topics such as foundational Buddhist teachings (e.g., *duhkha*, *anitya*, and *anatman*) in the introductory class, and Madhyamaka philosophy, epistemology, and debate methodologies in the advanced class. Student questions were only allowed at the end of class periods.

One day in the advanced class, Rinpoche focused on the relationship between the five *skandhas* (i.e., the five aggregates: matter, sensation, perception, mental formations, consciousness) and *sunyata*, emptiness. The lanky monk sat at the head of the classroom in his usual dark-red robes. After the introductory verses of dedication read aloud by the class, the monk remained quiet. Jim was there, too, wearing that same collared navy-blue shirt; he was also quiet and smiling. The still-steaming cup of tea, prepared earlier by the students, sat in front of the monk. The teacher kept his gaze directed at the cup as more moments passed by in silence. Long quiet lapses like this were common in Rinpoche's classroom. The monk then broke the silence, saying, "So, this cup." Another minute of silence passed by, with the monk and everyone else just looking at the cup. Jim kept smiling and looking around the room. Then Rinpoche said, "The cup is green only because I see it as green. It is not green all by itself, from its own side. So, it is empty."

A connection to the biological phenomenon of color-blindness sparked in my mind. Though I usually maintained a respectful student-learner mindset and demeanor in Rinpoche's classroom and kept my science mind at bay, in this instance I could not resist. I raised my hand in that moment, forgetting the rule that questions needed to wait until the end of class. The monk looked up from the cup and then into my eyes for what seemed like an eternity. My fellow students' gazes also turned away from the cup and toward my raised hand. I looked over at Jim; his smile was now bigger and directed at me. I considered putting my hand back down as seconds ticked by but felt that would be even more embarrassing. Finally, Rinpoche broke the silence with a simple, "Yes?"

I told the monk and the class that about 10 percent of the men, but far fewer women for genetic reasons, on the planet were color-blind and would actually see the cup as some kind of muted yellow rather than green. The monk leaned forward in his chair and an ever-slight hint of a smile appeared on his face. Another student spoke up, noting that he, in fact, was color-blind and always had to "fake it" when other people talked about colors. Rinpoche sat back in his chair, and his smile widened. My spontaneous, rude interjection at the beginning of the lecture had thankfully transformed into a welcome diversion. Jim asked why it was so common in men, but not women. I provided the answer, found in virtually all genetics textbooks, about the gene encoding a protein necessary for seeing color being present on the X chromosome, and that people who identify as women typically have two X chromosomes,[38] whereas those identifying as men typically have only one X chromosome, paired with a Y chromosome that doesn't have that gene. So, if a defective version of that gene occurs on one X chromosome in a person with two X chromosomes, a "good" copy of the gene on the other X chromosome can compensate for the bad one. Individuals carrying one X chromosome and one Y chromosome do not have such an opportunity for "genetic masking."

The genetic conversation continued and connected back to the Buddhist paradigm of the five *skandhas* and their relationship to *sunyata*, emptiness. Lots of students chimed in. Special focus centered on the transition from the second *skandha*, sensation, to the third *skandha*, perception, and the role that color-blindness might play in that. Rinpoche asked me a few questions, and I felt very, very honored.

This unexpected classroom conversation inspired me in new ways. It transformed how I thought about all kind of biological processes and entities, including DNA, the molecule at the center of my scientific attentions over the last two decades. I realized that the "DNA" in my mind was a function of the methods I used to detect and study it, and the impact that it has had on other aspects of my life. The "DNA" for me was different than the "DNA" for everyone else. The "DNA" conceived by the version of me after the Maitripa College classroom conversation was different than the "DNA" comprehended by the previous version of me. As Rinpoche returned to his original lesson plan, I continued to ponder the extent of my blindness to other wavelengths along the *sunyata* spectrum.

5

Intersections II

Theravada and Mahayana share core values and teachings originating from the Buddha (such as the Four Noble Truths), though the vast geography of Asia has limited interaction between these "southern" and "northern" Buddhist branches over much of the last two millennia. Theravada-Mahayana intersections became more prevalent over the last century with the advent of airplanes, the Internet, and other advances that made global travel and communication more feasible. Nonetheless, the many centuries of separation provided a historical context whereby Mahayana and Theravada Buddhists more or less independently intersected with scientists, offering the potential for both convergent conversations and distinct dialogues.

Buddhism spread across Asia, diversifying and coevolving with the traditions and religions encountered across the centuries. The northward path of Buddhism into Tibet gave rise to Madhyamaka philosophical powerhouses such as Tsongkhapa and the iconic Dalai Lama lineage of the Gelug School of Tibetan Buddhism. Another one of Buddhism's most important journeys was its eastward entry into China, where it flourished and adapted to already established cultures and traditions. Chinese philosophy is often described in terms of harmony among the "three teachings": those of Confucius (founder of Confucianism, ~500 BCE), Laozi (founder of Taoism, ~400 BCE),[1] and the Buddha.[2] One of the earliest Buddhist monks to enter China was An Shih-kao in 148 CE, who translated (from Sanskrit to Chinese) Theravada *suttas* found in the Pali Canon that explained yogic and Buddhist meditative practices.[3] Mahayana Buddhism, however, is recognized as having the strongest influence on the development of Buddhist traditions in China, which then spread further east into places such as Korea and Japan, where the Buddha's teachings once again encountered existing cultures with which it continued to coevolve.

Stories from Japan

Japan's first recorded exposure to Buddhism was in 467 CE, when five monks from the Gandhara region of northwest India visited the eastern islands.[4] However, the "official" introduction of Buddhism to Japan is dated to 552 CE, when King Seong of Baekje (what is now western Korea) sent a Buddhist mission to Japan that included *bhiksus* and *bhiksunis* (Sanskrit; *bhikkus* and *bhikkunis* in Pali) who arrived with images of the Buddha, and Mahayana *sutras*. The Buddha's teachings subsequently spread across Japan over the next ~1,500 years, with the development of many diverse Zen, Pure Land, Shingon, and other Mahayana schools.

During the latter half of the nineteenth century in Japan, much like the situation in Sri Lanka, Buddhist culture and its thinkers were confronted by social and spiritual challenges from Europe, including Christianity, evolutionary science, and the ongoing conflict between these two Western worldviews. For example, Inoue Enryo, a late nineteenth-/early twentieth-century Pure Land Buddhist priest from Japan, was a world traveler, prolific writer, and an early advocate of evolutionary theory in Japan.[5] Enryo embraced Darwinian evolution and actively deployed it as a tool to counter the influx of foreign Christian concepts (e.g., creationism) into Japanese culture. The Mahayana Buddhist priest rejected the notion that there was an essential, inherent distinction between humans and animals, but rather a gradual one that was in line with evolutionary concepts. The writings and speeches of Enryo contributed to a Japanese culture conducive to the further engagement of Buddhist priests in the Buddhism and science dialogue.

The 1893 World Parliament of Religions in Chicago hosted numerous representatives from the many forms of Buddhism, such as Anagarika Dharmapala from Sri Lanka. Although Inoue Enryo was not in attendance at this global gathering, other delegates from Japan were there and argued that Buddhism harmonized with modern science. Shaku Soen, a Rinzai Zen priest and Mahayana master, attended the parliament and emphasized the science-compatible, Buddhist core concepts of impermanence (*anitya*), suffering/dissatisfaction (*duhkha*), and cause and effect (*pratityasamutpada*) in his address:[6]

As the phenomena of the external world are various and marvelous, so is the internal attitude of the human mind. Shall we ask for the explanation of these marvelous phenomena? Why is the universe in constant flux? Why do things change? Why is the mind subjected to a constant agitation? For these Buddhism offers only one explanation, namely the law of cause and effect.

Doki Horyu, a second Japanese delegate to the 1893 World Parliament of Religions, represented Shingon ("mantra") Buddhism. Shingon constituted a lineage of Vajrayana (also referred to as and "tantric"), which heavily influenced the development of Buddhism in north-central regions of Asia such as Bhutan, Mongolia, and Tibet. Shingon and other Vajrayana lineages offered distinctive and esoteric practices and philosophies as compared to the more prominent Mahayana schools (e.g., Zen and Pure Land varieties) of Japan. After the parliament in Chicago, Horyu traveled widely through Europe and India before returning home to Japan. The Buddhist priest visited London, where he encountered a fellow Shingon Buddhist from Japan, Minakata Kumagusu, on October 30, 1893.[7] Kumagusu, residing in London at the time, spent several days in discourse with the Shingon priest, which transformed into a lasting relationship of written correspondence across many years. The letters between these two Shingon Buddhists were rediscovered in the 1970s and brought forth to the world the unique thoughts of Kumagusa: a reclusive Japanese scholar who published fifty articles in *Nature* and sought to create a new Buddhist paradigm for science.[8]

* * *

Minakata Kumagusu was born into a devout Shingon family in the city of Wakayama. As an infant, his parents brought him to the Fujishiro shrine and a resident priest gave him the name "Kumagusu" by combining the Japanese characters for *kuma* (bear) and *kusu* (camphor tree). As a child, Kumagusu read deep significance into this name; for him, it indicated a fundamental interconnection between animals, plants, and himself. Kumagusu initially pursued higher learning at the University of Tokyo, though he dropped out at the age of nineteen to pursue studies in America. He studied at the Michigan State School of Agriculture (now Michigan State University), but he was expelled after a heavy drinking incident with some other students. Kumagusu then left the United States and journeyed across Cuba and South America with a traveling circus before crossing the Atlantic Ocean and arriving in England in 1892. Kumagusu spent the next eight years in England, where he secured a position as an Oriental Research Fellow at the British Museum. During this time, Kumagusu met the aforementioned Shingon priest Doki Horyu, with whom he developed a strong personal relationship. He also interacted with a variety of naturalists and other European scholars at the museum. Kumagusu discovered the writings of Charles Darwin, Herbert Spencer, and other prominent European biological thinkers of the time. In 1893, Kumagusu published his first article for *Nature*, "The Constellations of the Far East,"[9] which attracted praise in British scientific society. Kumagusu continued to publish in *Nature* and other journals on a diversity of topics ranging from biology to folklore to sexology. In 1899, however, an incident resulted in the expulsion of the young Japanese man from white academic society once again. The episode began with a conversation between Kumagusu and a "Mr. Thompson," whereby the latter individual was lecturing Kumagusu about the inferiority of Orientals relative to Europeans. Minakata Kumagusu's response was to punch Mr. Thompson in the face, right there in front of hundreds of people in the museum reading room. This infamous incident resulted in Kumagusu's exile from the British Museum and his eventual return to Japan.

Kumagusu received a frosty reception upon his return home. News of the incident at the British Museum had made its way back to Japan. Further, though a remarkably well-published and prolific scholar, he had failed to secure an academic degree during his many years abroad in America and Europe. Kumagusu had returned to his home in disgrace.

In 1904, the Japanese scholar set out on a mission of biological discovery and solitude. He took residence at an inn by a small Shingon shrine, near the town of Nachi. Over the next two and a half years, Kumagusu repeatedly ventured into the mountainous forests surrounding the shrine by day. He collected and observed the behaviors of fungi, slime molds, and other peculiar organisms that he frequently encountered in the woods. By night, Kumagusa analyzed his specimens using a microscope brought back from England. He also read. Kumagusu drew inspiration from Buddhist writings that included the Mahayana *Kegon Sutra* (Flower Garland *Sutra*) and the *Treatise on the Five Teachings of the Huayan* by Fazan, a seventh-century monk from China. *Pratityasamutpada* (dependent arising) was a prominent feature of both of these writings, expressed through the metaphor of Indra's net, introduced earlier in Chapter 2. Indra's net represents the mutually interdependent nature of the universe, with multifaceted jewels present at each intersection that reflect the images of all the other jewels in the net. All jewels, elements of the universe, are simultaneously reflecting and reflected in infinite directions.

Slime molds fascinated Kumagusu. Like his name, they expressed some features consistent with animal forms of life, and others more similar to plants. These organisms underwent profound morphological changes during their development process, which captivated the Japanese Buddhist-biologist recluse. In short, the slime mold life cycle involved an initial phase of microscopic single-celled spores that moved and behaved as individuals. In a second stage, thousands of earlier independent spores coordinately fused to form a larger plasmodium, commonly called the "slug" phase, which was either a single cell with multiple nuclei or multiple cells that lack membrane separation, depending on one's perspective. The slime-mold slugs were visible to the naked eye and mobile; Kumagusu would observe them moving on dead and decaying trees, leaving a "slime" track in their wake. In a third stage, the slugs immobilized and transformed into stationary fruiting bodies that exhibited a wide array of varying shapes and colors; the fruiting bodies harbored the microscopic single-celled spores capable of beginning a new cycle.

The ambiguous and dynamic nature of slime molds posed a vexing and frustrating problem for Europeans biologists and taxonomists. Its features did not fit into their system of classification. Kumagusu, however, reveled in the fact that these organisms defied the taxonomic paradigm of European scientists. The slime mold research led Kumagusu to conclude that, because these organisms had no fixed "animal" or "plant" nature and displayed dynamic changes across different life stages, the overall taxonomic classification approach of European science had limited overall value.[10] Kumagusu discovered ninety-nine slime mold species during his life, and a slime mold genus, *Minakatella*, was named after him to honor his contributions to this understudied corner of biology.

Later in life, Kumagusu abandoned his existence of isolation and moved to the town of Tanabe; there he married the daughter of a local Shingon priest. His life took on a more domesticated path. Kumagusu's slime mold research, however, continued:[11]

> I have patiently observed for ten to twenty years, and when thus observing the same [organism] in the same place, even in a small place like [my garden] in Tanabe, [one can see that] in this way, today, too, without human help, organisms naturally change into new species, or are changing and unstable. In my old house, they become new species, while in my new house, they return in the opposite direction to their original stock and so on; hence, one can see that in this wide universe, without the help of human intervention, there are constantly innumerable changes occurring.

In Tanabe, Kumagusu's activities also pivoted toward environmental activism. The local Shingon shrine led by his wife's family was under threat from a governmental effort to consolidate the many thousands of small, local Buddhist shrines around Japan into a smaller number of large, government-sanctioned temples. Kumagusu demonstrated against this effort through speeches and the production of political pamphlets, arguing about the long-term negative impacts the consolidation effort would have on the society and the ecology of Japan. The thousands of small shrines around Japan were often adjacent and served to protect surrounding forests, which held significant spiritual value to the local communities and supported the slime molds and other life forms held dear by Kumagusu. Elimination of these sacred shrines, however, opened up the nearby forests to logging and other economic development initiatives. Kumagusu's efforts eventually curtailed the governmental efforts of shrine amalgamation, though not before thousands had been destroyed. In 1910, Kumagusu had another alcohol-fueled incident whereby the inebriated Buddhist biologist stormed into and disrupted a community gathering organized by government officials to promote the temple consolidation effort. He made quite a scene. Though this drunk and disorderly incident landed Kumagusu in jail, the scholar serendipitously discovered a new species of slime mold inhabiting his prison cell.[12]

For many years, Kumagusu remained a mostly unknown oddity, characterized as an eccentric though simultaneously unimaginative scholar. He was caricaturized as an uncreative "collector" of slime molds, who didn't publish most of his biological research, drank too much, and spent a lot of time wandering in the woods. A different picture emerged, however, when a 1970s sociologist named Kazuko Tsurumi discovered Kumagusu's lengthy written correspondences with Doki Horyu, the Shingon monk he initially met in London during his time at the British Museum. Tsurumi, a Tokyo-born and Ivy League–educated scholar, brought forth the lengthy correspondence between Minakata Kumagusu and Doki Horyo. Tsurumi's writing portrayed Kumagusu, the slime mold aficionado, to be a much deeper, bolder, and creative thinker than previous characterizations of this scholar in the biological sciences community. The sociologist framed Kumagusu as a pioneering thinker on Buddhism-science intersections and highlighted a ten-meter-long 1903 letter written from Kumagusu to Horyo, which discussed matters such as the limits of linear thinking and criticized the basic models of cause and effect that underpinned Western science. In the letter, said to have resulted from three days and nights of sleepless intellectual frenzy, Kumagusu also presented diagrams illustrating a new framework for cause and effect based on *pratityasamutpada*, the central topic of the Buddhist readings the recluse brought into the woods, and popularized by Kazuko as the "Minakata Mandala" (see Figure 5.1).

The term "mandala" describes a variety of geometric designs and artistic expressions used in Buddhist and other Asian traditions. Mandalas are prominent in Vajrayana forms of Buddhism, such as the well-known sand mandalas constructed by teams of Tibetan monks that depict Mahayana deities and *bodhisattvas*, and express core Buddhist concepts such as *anitya* (impermanence). In Japanese Shingon Buddhism, one prominent mandala portrays Vairocana, a celestial Buddha who embodies the *sunyata* (emptiness) concept, encircled by eight supporting *bodhisattvas*. The Manakata Mandala, named by the Japanese Buddhist philosopher Nakamura Hajime, depicts an array of intersecting lines,

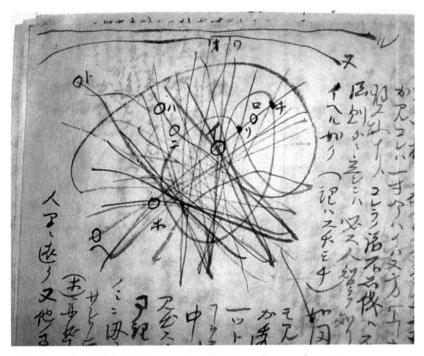

Figure 5.1 Minakata Mandala. Originally drawn by Minakata Kumagusu in a letter to Doki Horyu on July 18, 1903. Copyright reserved, Minakata Kumagusu Archives (Tanabe City).

sometimes referred to as a "spilled bowl of noodles."[13] In the diagram, in-dividual lines represent particular chains of cause and effect; line paths are affected by the trajectories of nearby lines in the mandala. Dense concentrations of intersecting lines appear in the mandala, which Kazuko Tsurumi and others interpreted as representing multiple cause-and-ef-fect vectors coming together under observation by a human being's mind. Other intersections on the outskirts of the diagram represented intersecting vectors that went unobserved by humans, representing the limits of human perception. Kazuko argued that the Manakata Mandala represented the start of an entirely new paradigm for science rooted in Buddhism's mutual cause-and-effect framework of *pratityasamutpada*. She saw it as a radical potential paradigm shift for science, moving away from the Eurocentric, linear cause-and-effect framework of Newtonian logic and toward a new model of mutual interdependent causes and effects, rooted in the teachings of the Buddha. This bold vision, however, never quite took off, at least in the world of science. This is not surprising, however, given the entrenched structural disconnections between social scientists and biologists in Western academia, and the broader temporal and geographic separation of Buddhism and science over the last two and a half millennia. Today, Minakata Kumagusu remains a mostly unknown entity in the world of Western academic biological research.

The Dalai Lama Takes on Mutation

Tenzin Gyatso, His Holiness XIV Dalai Lama of the Gelug School of Tibetan Buddhism, is a world-famous monk largely responsible for popularizing the parallels between Buddhism and science.[14] The Dalai Lama's fascination with science started when he was a young boy; he tinkered with watches, telescopes, and a rare automobile at the Potala Palace of Tibet. His Mind and Life Dialogues, initiated with the late great philosopher of biology Francisco Valera in 1983, brought together eminent scientists, philosophers, and other scholars to engage in deep dialogues with the monk. The Dalai Lama has frequently supported scientific views on issues ranging from quantum physics to psychology. The superstar Tibetan monk famously stated, "If sci-entific analysis were conclusively to demonstrate certain claims in Buddhism to be false, then we must accept the findings of science and abandon those claims."[15] The Dalai Lama, however, continues to accept and defend the

Buddhist concept of human rebirth, a favorite target of scrutiny by Western scientific thinkers.[16]

Each of the Dalai Lama's Mind and Life Dialogues focused on one or a few specific themes; the 2002 gathering held in Dharamshala, India, was titled "The Nature of Matter, The Nature of Life." The second half of the event focused on questions of biological complexity and evolution, and featured renowned geneticists such as Eric Lander, Massachusetts Institute of Technology professor and a driving force behind the human genome project, and Ursula Goodenough, a Washington University professor and author of *The Sacred Depths of Nature*.[17] Eminent Western Buddhist thinkers also attended, such as Matthieu Ricard, a French-born molecular biologist-turned-Buddhist monk.

During the dialogue on evolution, the Dalai Lama saw extensive harmony between his Buddhist mindset and the characterizations of Darwinian evolution and other aspects of biology presented by the scientists, though with a few key exceptions. The process of DNA mutation, the underlying "fuel" for evolution and substrate upon which natural selection acts according to modern evolutionary theory, was characterized as an entirely random process by the attending scientists. This did not sit well with the monk. The Dalai Lama commented:[18]

> From a philosophical point of view, the idea that these mutations, which have such far-reaching implications, take place naturally is unproblematic, but that they are purely random strikes me as unsatisfying. It leaves open the question of whether this randomness is best understood as an objective feature of reality or better understood as indicating some kind of hidden causality.

The Dalai Lama doubted the proposition that DNA changes in a random, willy-nilly fashion, free from the broader context of cause and effect expressed in Buddhist thought (i.e., in accord with *pratityasamutpada*). The idea that mutation was totally random, with any particular piece of DNA being just as likely to mutate as any other regardless of prior experiences or interactions, opposed Buddhism's view of how things work. Can a biological process so central to human disease and life's evolution truly be "random"?

The Dalai Lama also found fault, a sense of incompleteness, in the theory of natural selection because the reproductive competition it postulated seemed to leave out the possibility for altruism, a feature widely observed in the natural world and of central value in the Buddha's teachings. This apparent evolutionary aspect has become a target of scrutiny by other Western scientists, such as David Sloan Wilson[19] and Joan Roughgarden,[20] as well as other Buddhist thinkers, such as Inoue Enryo[21] and Albert Low,[22] a Zen teacher based in Montreal. These two key criticisms of evolution by the Dalai Lama—that mutation is purely random and natural selection precludes the possibility of altruism—will be the focus of further analysis in later chapters.

Evolutionary Eyes

Encounters between biologists and Buddhism have increased in frequency over the last twenty years. Academic evolutionists entering the Buddhism-science stream often orient their discourse in the context of central questions and themes in the biological sciences, such as the famous "selfish gene" hypothesis posed by Richard Dawkins in the late twentieth century. Dawkins's perspective in *The Selfish Gene*[23] presented a world in which genes encoded on the DNA molecule were the central drivers of all other biological phenomena, and that the organisms which house them were mere "machines" constructed by the genes. In Dawkins's well-known words:[24]

> What weird engines of self-preservation would the millennia bring forth? Four thousand million years on, what was to be the fate of the ancient replicators? They did not die out, for they are past masters of the survival arts. But do not look for them floating loose in the sea; they gave up that cavalier freedom long ago. Now they swarm in huge colonies, safe inside gigantic lumbering robots, sealed off from the outside world, communicating with it by tortuous indirect routes, manipulating it by remote control. They are in you and in me; they created us, body and mind; and their preservation is the ultimate rationale for our existence. They have come a long way, those replicators. Now they go by the name of genes, and we are their survival machines.

Although this gene-centric view of evolutionary biology, often referred to as "neo-Darwinism," became a prominent shaper of evolutionary thought in academic circles and the public sphere, the theory also became a target for criticism by many biologists and philosophers of science. Many of those biologists levying criticism of Dawkins's reductionist, gene-centric view of evolutionary biology also discovered that their views align with Buddhist teachings.

* * *

Denis Noble, an English-born emeritus professor of Oxford University who in 1960 developed the first successful mathematical model simulating the living human heart,[25] organized the first debate on Dawkins's selfish gene concept in 1976 and has published more than fifty articles on the subject. Noble's most detailed critique of the selfish gene theory and, more broadly, the "central dogma" of molecular biology that preceded and underpinned it, appeared in *The Music of Life: Biology Beyond Genes*.[26] In this book, Noble outlined his then-new theory of systems biology, which argued against the primacy or privilege of DNA, or any other biological entity (e.g., cell, tissue, organism) in the functioning of biological systems. The Oxford biologist rejected the notion that DNA, all by itself, provided any kind of genetic "program,"[27] but rather served as a "database" in biological systems. Noble further de-emphasized the relevance of any particular specific biological entity in and of itself (e.g., DNA, protein, cell, etc.) and instead pointed to the interactions among these factors as the keys to understanding how life works. Noble noted the myopic and misleading nature of the unidirectional and "bottom-up" model of the central dogma of molecular biology (i.e., DNA → RNA → protein). He referred to this as the "reductionist causal chain" and argued against it with many biological examples where "higher" level factors influence those at "lower" levels (e.g., the replication of DNA by protein machinery, organ-level effects on gene expression mediated by hormones). After nine chapters dedicated to a systematic breaking down of the reductionist causal chain and building up of his systems biology theory, the tenth and final chapter of *The Music of Life*, "Curtain Call: The Artist Disappears," Noble explicitly tied his systems theory to Buddhism's concept of *anatman* (no-self). He wrote:[28]

> If the "self," the "I" as Descartes conceived it, and in new guise as modern neuroscience tends to conceive it, is an object that we hang on to because our language and culture make it very difficult to do otherwise, then clearly it is important to know that there are cultures in which it doesn't exist—or at least in which it doesn't exist in these senses, neither as a separate substance (the Cartesian view) interacting with the brain, or (the modern view) as part of the brain itself.

For 2500 years this has been part of the aim of Buddhist meditation. There are many forms of Buddhism around the world, with a wide range of practices and beliefs, but the idea of "selflessness," "the disappearing self," and "letting go" is common. In some forms, there is little or no metaphysics either, just a code of practice: a religion, one might say, without beliefs. And, as such, it contains no possibility of conflict with science.

Noble more thoroughly expounded on the relationships between Buddhism and his systems biology theory at two events in 2010: a colloquium on Buddhism and Science at Oxford University, and at the 3rd World Conference on Buddhism and Science held at Mahidol University in Thailand. For the latter event, Noble presented a paper,[29] "Convergence between Buddhism and Science: Systems Biology and the Concept of No-Self (*anatman*)," which presented an overview of his systems biology theory, followed by comparisons to Buddhist philosophy and practice. The Oxford professor focused on the convergent nature by which he and Buddhism arrived at a similar insight (that of *anatman*), but he emphasized the distinct methodologies by which those conclusions arrived. Noble's many decades of research and reflection on biological science led him to reach a "no-self" conclusion versus the meditative direct-personal experience route of Buddhists. Noble framed his version of "no-self" as a conceptual insight and the "no-self" of Buddhism as an experiential fact.

East-Asian forms of Mahayana Buddhism were prevalent in Noble's writings and talks. In *The Music of Life*, Noble highlighted the ox herder story, a popular parable among medieval and contemporary Japanese Zen Buddhists, which likened a herdsman's search for his missing ox to an individual's journey toward enlightenment. In his 2010 paper presented in Thailand, the Oxford biologist shared a writing composed by Won Hyo, a seventh-century CE Korean monk and commentator on Mahayana *sutras*, that described the interrelationships between a seed and the fruit to illustrate Buddhist views on being/non-being, and the Mahayana Middle Way. Noble modified Won Hyo's prose, replacing the word "seed" with "genotype" (a genetic term used to describe the genetic "type" of an organism) and the word "fruit" with "phenotype" (the set of observable characteristics of an organism, resulting from its genotype and interactions with the environment). The Oxford professor's modified version[30] of the Korean monk's writing served to illustrate his systems biology view:

The phenotype and the genotype are not the same, for they have different shape.

However, they are not different.
Besides the genotype and the phenotype are not annihilable,
for the phenotype is produced from the genotype.
However, they are not eternal,
for there is no genotype when it is in the state of the phenotype.
The genotype did not enter into the phenotype,
for the genotype does not exist when it is in the state of the phenotype.
The phenotype does not extinguish the genotype,
for the phenotype does not exist when it is in the state of the genotype.
Since it neither enters nor is extinguished, there is no arising.
Since it is neither eternal nor annihilable, there is no ceasing.
Since there is no ceasing, non-being cannot be proclaimed.
Since there is no arising, being cannot be proclaimed.

* * *

David Barash, another Western evolutionary thinker, waded into Buddhist waters. Barash, emeritus professor at the University of Washington, first published his early thoughts on Buddhism-biology parallels in a 1973 article titled "The Ecologist as Zen Master."[31] After four decades of further contemplation and conceptual development, Barash published a more thorough analysis: *Buddhist Biology: Ancient Eastern Wisdom Meets Modern Western Science.*[32] He examined the ways by which the Buddha's teachings harmonize with scientific views, with special emphasis on ecology and evolutionary biology. Barash oriented his outlook with a comparison to the thinking of the late great Harvard evolutionary biologist, Stephen Jay Gould. On the question of the relationship between science and religion, Gould famously advocated the principle of "non-overlapping magesteria" (NOMA),[33] which proposed that each domain represented distinct modes of inquiry, such that there was a difference between the "nets" over which they had a legitimate magisterium, or domain of authority. In short, science and religion should not overlap, according to the NOMA doctrine. *Buddhist Biology*, by contrast, argued that in the interaction between Buddhism and science there is strong potential for "productively overlapping magisteria" (POMA). David Barash, like Denis Noble, found convergence in the science and Buddhism's conclusion of *anatman,* though the former scientist framed this conclusion more in an environmental and ecological context as compared to the molecular genetic view of the Oxford professor:[34]

> As Buddhist masters have known for thousands of years, and "biology masters" have been maintaining for considerably less, the human skin does not separate us from the rest of our environment—it joins us to it. And the more joining there is, the less clear are the boundaries between us and the rest of the world, until it becomes evident that there is no "us" inside distinct from the "them" or "it" outside.

The Washington professor, however, went further to connect biology to two other fundamental Buddhist principles, *anitya* (impermanence) and *pratityasamutpada* (dependent arising). Barash highlighted the limits of overly simple cause-and-effect relationships often deployed in science and the interconnectedness of everything in nature. Barash made frequent reference to Thich Nhat Hanh, an influential Vietnamese Zen monk who popularized the "interbeing" conceptualization of *pratityasamutpada* that underpins his brand of "Engaged Buddhism."[35] *Buddhist Biology* also looked at reductionist approaches, such as the selfish gene theory, versus holistic approaches that aim to tackle problems of complexity and proposed *pratitysamutpada* as a potential middle-ground strategy to reconciling "greedy reductionism" versus "lazy holism." Barash's book concluded on a philosophical note and made connections to Western philosophy, beckoning further work at the trisection of Buddhism, biology, and existentialism.[36]

More evolutionary eyes have turned to Buddhist ideas in the last two decades. In 2002, Sydney Brenner, a recently deceased Nobel laureate who made seminal contributions to the development of the central dogma of molecular biology and pioneered the use of the worm *Caenorhabditis elegans* in genetic analysis, wrote an article, "Elementary Zenetics; do or dai."[37] This article recounted Brenner's visit to Japan, where he listened to a lecture by Susumu Ohno, a Japanese evolutionary geneticist famous for analyzing and spotlighting the relevance of duplicated genes in evolution. Ohno's presentation incorporated musical translations of DNA sequences. This experience inspired Brenner to think of new "Zen-inspired" ways of looking at genes through Zen gardens, *haiku,* and *koans.*

In 2017, Scott Gilbert, emeritus professor of biology at Swarthmore College and the University of Helsinki, spoke at the first congress on Science and Tibetan Buddhism, held in conjunction with six hundredth anniversary of the Drepung Monastery in India. The event saw the first group of twenty nuns receive the *geshe* degree (equivalent to a PhD) of Gelug Tibetan Buddhism,[38] and it was attended by the Dalai Lama. In his conversation with the Dalai Lama, Gilbert highlighted convergence between biology and Buddhism by referencing parallels between *anatman* and the recently discovered role of bacteria in shaping animal developmental processes. Gilbert recounted his lecture on human developmental biology presented to the Dalai Lama, and its connections to Buddhist principles, in the *Swarthmore College Bulletin*:

> My talk focused on the critical importance of reciprocal interactions during human development. This includes the cooperation between two cells during fertilization, the cooperation between two tissues during organ formation, and the cooperation between several species (bacteria and mammalian) to generate gut capillaries shortly after birth. It linked these to the fundamental Buddhist notion of dependent co-origination, the idea that nothing has its own essence, but rather is formed in concert with other things.

Fred Allendorf, emeritus professor of biological sciences at the University of Montana, is another example of an evolutionary ecologist who has discovered parallels with the Buddha's teachings. This conservation-minded biologist has studied the migration and flows of salmon, and the DNA inside them, in the Pacific Northwest of the United States throughout his academic career. In 1997, Allendorf published his first essay at the Buddhism-biology interface, "The Conservation Biologist as Zen student,"[39] which recognized and started to build upon David Barash's pioneering 1973 article about Zen and ecology. The evolutionary ecologist recently expanded his thoughts on the Zen-biology interface with his 2018 article "Zen and Deep Evolution: The Optical Delusion of Separation."[40] In this piece, Allendorf addressed criticisms of the Buddhism-biology "POMA" possibility, levied by Jerry Coyne in *Faith Versus Fact: Why Science and Religion Sre Incompatible*,[41] related to Buddhist beliefs in reincarnation and karma, and the previously discussed concerns of the Dalai Lama relating to the claims of randomness in the spontaneous mutation process. The Montana fish biologist supported Barash's POMA view by highlighting Thich Nhat Hanh's secularized views on reincarnation and arguing for compatibility between the role of chance in mutation and Buddhist notions of cause and effect (i.e., "karma," *pratitysamutpada*). Allendorf's essay made connections between Buddhism and Moto Kimura's influential neutral theory of molecular evolution,[42] and it emphasized experiential Buddhist meditative practice by providing a guided meditation focused on the carbon cycle of ecosystem ecology. Allendorf deployed the ethical framework of Buddhism as a motivational tool to promote environmental compassion and action, with special focus on the extinction crisis.

In 2006, the Emory-Tibet Science Initiative was launched; this remarkable and ongoing effort was intentionally designed to teach modern science to Tibetan monastics, and it was centered in the biological sciences. Arri Eisen, an Emory Professor of Biology, and Yungdrung Konchok, a monastic participant in the Emory-Science Tibet Initiative, coauthored *The Enlightened Gene: Biology, Buddhism, and the Convergence That Explains the World*.[43] This book tackled a multitude of diverse biological topics and perspectives ranging from the nature of sentience to ecology and karma to meditation and stress prevention. The fourth chapter, "Altitude and Attitude," discussed the remarkable mental and physical adaptations of Tibetan people to their Himalayan environment which revealed key insights into DNA–environment interactions. The authors emphasized how the long-term and deliberate synthesis of biology and Buddhism through the Emory-Tibet Science Initiative changed their perspectives on the nature of nature and the rest of the universe.

David Sloan Wilson, professor of biology and anthropology at Binghamton University in New York and a cofounder of the Evolution Institute, has focused much of his research career on investigating and legitimizing the process group selection in evolutionary processes as an alternative to Dawkins's neo-Darwinian "selfish gene" theory. Wilson's work makes connections between social and cultural evolution among human beings. In his recent book, *This View of Life: Completing the Darwinian Revolution*,[44] Wilson emphasized the need for humans to recognize and embrace their active role as "wise managers of evolutionary processes." The New York professor built upon previous analyses of altruism,[45] demonstrating how behaviors such as goodness and altruism can evolve under the right evolutionary circumstances. Wilson met with the Dalai Lama at the 2019 Mind and Life Dialogue, where the two discussed common goals in developing ethical frameworks that can transcend religious, cultural, and racial differences.

* * *

The seven previously discussed Western biologists encountered and worked productively with Buddhism in a variety of ways, but they also remained scientists at the core. The eighth and final example to be discussed, Mathieu Ricard, abandoned his life of science to become a Buddhist monk. Ricard grew up among French intellectuals, raised by a father who was a well-known secularist philosopher and writer, and a stalwart champion of science. In the late 1960s and early seventies, Ricard was a PhD student at the Institut Pasteur in Paris, under the supervision of famed Nobel Laureate Francois Jacob whose research revealed foundational insights into the interactive relationships between DNA, RNA, and proteins. As a graduate student in the Jacob lab, Ricard investigated the function of DNA in the model bacteria *Escherichia coli*, similar to most of his scientific contemporaries, and ultimately earned a PhD for his work. Ricard also visited India on occasion through the course of his graduate studies and there became exposed to Tibetan Buddhism. Ricard's graduating peers had their eyes fixed on the next biological research frontiers that went beyond "simple" bacterial systems, such as investigating the genetic underpinnings of more complex systems such as animal cells and development, neuronal patterning, and cancer. Ricard took a different path.

After receiving the PhD in 1972, Ricard completely abandoned science and moved to India, where he began study under a series of Tibetan Buddhist teachers, starting with Kangyur Rinpoche in Darjeeling. This new path eventually led Ricard to trade his white lab coat for the burgundy robes of a Tibetan Buddhist monk. After his ordination as a monk, Ricard transitioned to a monastery in Nepal, where he now coordinates humanitarian projects, translates Tibetan texts into other languages, and writes books. In his work, Ricard directly confronts and shares his rationales for transitioning to a monastic life and adopting the teachings of the Buddha. In *The Quantum and the Lotus*,[46] for example, Ricard and University of Virginia astronomer Trinh Xuan Thuan share dialogue that compares and contrasts scientific and Buddhist notions on "being" and the basic nature of reality.

Ricard has fully embraced all aspects of the Tibetan Buddhist path, including an acceptance of human rebirth. How does Ricard, an individual raised in the rational waters of European secularism and trained at the highest level in genetics, rationalize rebirth? He explained this in *The Monk and The Philosopher*, a book that Ricard coauthored with his father. The monk explained that Buddhism holds that there exists a nonmaterial component to consciousness, and that this feature flows like rivers through time and space. Ricard framed a rationale for rebirth through Tibetan Buddhism's three criteria for valid knowledge: direct experience (or perception), irrefutable deduction (or inference), and testimony worthy of confidence. Ricard recounted that trustworthy teachers, never known to lie or be deceptive, claimed the ability to perceive streams of consciousness. Many such teachers through time, going back to the Buddha, claimed direct knowledge of past lives gained through meditative or other experiences. For Ricard and most others who hold rebirth as a valid truth, the third criterion for valid knowledge is essential. Trustworthy beings should simply be trusted. Most secular science minds, however, are certainly unlikely to embrace Ricard's rationalization because it presents a knowledge framework that is refractory to outside evaluation; no one can "be there with you" while directly experiencing evidence of rebirth during meditation. Ricard and other Buddhists widely recognize that their evidence and rationales for human rebirth are not generally deemed as acceptable from scientific eyes. However, they generally see this as being a problem relating to the limits of science rather than a problem for Buddhists.

Motivations and Privilege

What motivates and allows Buddhists to seek out biology? For Bhikkuni Kusuma, she ventured across oceans and left behind family to seek scientific answers to the fundamental questions of life in America, but ultimately became disillusioned and returned home to Sri Lanka and became a Buddhist nun; there she found her answers. Likewise, the Dalai Lama's Mind and Life Dialogues seek answers to big questions, and bring together thinkers from varied traditions to collaboratively discuss and dissect deep topics, such as the nature of life. Minakata Kumagusu, the eccentric slime mold discoverer and environmental activism pioneer from Japan, attempted an all-out epistemic integration of Buddhism and science in his "Minakata Mandala" diagram. For Anagarika Dharmapala and Inoue Enryo, however, their narratives on the harmonies between evolutionary biology and Buddhism blended in sociopolitical motivations. Both of these Buddhists saw and wrote about synergies between Darwinian evolution and their respective brands of Buddhism, and they used this as a tool to combat the colonial Christian powers of Europe that threatened their native Asian island homes. The ability of these Buddhist thinkers to access the teachings of science derived in part from their positions of power. For example, the Dalai Lama had access to a telescope in the Potala Palace before he fled the Chinese invasion, a privilege unavailable to virtually everyone else in Tibet at the time. The other Buddhist thinkers discussed in this chapter also benefited from travel across the world to Europe and America, opportunities available to only a privileged few in colonial Asia.

What motivates and allows biologists to seek out Buddhism? In his introductory remarks at the 2010 Buddhism and Science Colloquium held at Oxford University,[47] Denis Noble commented, "I see two routes through which people like me have become involved in this kind of dialogue." He went on to first point to the convergence between his Western science-originating theory of systems biology and *anatman* (non-self), presented in the *The Music of Life*, and then secondly recognized the efforts of Buddhist monks and philosophers (such as the Dalai Lama and the conference organizers, Verna and Alan Wallace) in reaching out to scientists through Mind and Life conferences and similar efforts. David Barash, Fred Allendorf, and David Sloan Wilson highlighted connections to ethics, a feature frequently argued to be a significant void in evolutionary theories. These perspectives argued in favor of Buddhism and biology as POMA. For Sydney Brenner, Zen seems to have simply offered a vehicle for different ways of seeing DNA. For Mathieu Ricard, Buddhism was concluded to offer a superior vehicle relative to science for accessing truths, resulting in the scientist becoming a monk.

Evolutionary eyes are encased in skulls. These skulls are swathed in nerves, blood vessels, muscles, and lymph glands. All of this is wrapped in skin. For most evolutionary eyes, including my eyes, that skin is white. That white skin brings privilege in human societies. That privilege allows evolutionary eyes such as my eyes, and the brains to which they are connected, such as my brain, access to the wondrous waters of wisdom at the confluence of Buddhist and biological streams.

Life after Life

The 2012 fall semester at Maitripa College was wrapping up. The preceding months interacting with and learning from Yangsi Rinpoche, Jim Blumenthal, and the Maitripa students equipped me with a new Madhyamaka philosophical knowledge base and exposed me to meditative practices that would come to be essential to my personal well-being. On the final day of class, I arrived with my final essay in hand to turn in and found my usual desk spot in the lecture room. I sat next to a student who was currently working as a nurse but was at Maitripa pursuing a master's degree in divinity and wanted to transition to work in hospice. The bell rang, and everyone stood and bowed toward Rinpoche as he walked into the room one last time this semester. After the introductory chanting and other rituals were completed, Rinpoche looked at the class for a few minutes in silence and then looked at Jim, who, as usual, was seated to the monk's left. Jim was, of course, smiling and wearing that same navy-blue shirt. Jim then nodded in silence to the monk. Then Rinpoche said, "I have some big news. His Holiness the Dalai Lama will visit Portland and Maitripa College in March. There will be a small Buddhism and science symposium with His Holiness held in the *Jokhang* (meditation hall), and you are all invited." Rinpoche gave more details; he instructed that we would all need to purchase "Zen benches" because not all of the invited guests to this event would fit if we all sat on the usual meditation cushions. After a final short lesson focused on compassion, the monk rose and all the students stood up as well. For this final class session, the monk approached every single student, one by one, on the way to the door and pressed his nose against each student's nose for a few seconds. I was last; after our nose-to-nose encounter, Rinpoche smiled at me and then walked out the door.

* * *

The Dalai Lama visited Portland for an Environmental Summit during March 2013. He held a series of public speaking events that included a sold-out visit to the Rose Garden (now called Moda Center) sports arena where he gave a speech titled "Inspiration for the Global Environment." The Dalai Lama also visited Maitripa College on May 10, the day before the large Rose Garden event; for the small Buddhism and science symposium promised to the students of Maitripa College months earlier.

I was invited to attend the Buddhism and science symposium just as all the other Maitripa College students had been. An earlier version of me, such as the one that felt disrespected at the initial Maitripa visit to make sabbatical arrangements, might have felt slighted that I was a mere attendee rather than a featured panelist or something like that. By the time of Rinpoche's announcement at the end of the semester about the Dalai Lama's visit, however, I had become accustomed to a student-like rather than teacher-like identity in the Maitripa College environment. I was happy, excited, and humbled that I would get to listen to the Dalai Lama once again, nearly one decade since the words of the world-famous monk shocked my senses and shifted my sails in Bloomington, Indiana. Leading up to the event, my mind frequently engaged in speculative curiosity about what areas of Buddhism-science harmony might be explored in the symposium; given the broader environmental theme of the visit, I assumed it would be something related to ecology or the environment. But areas such as neuroscience and physics were also prominent science topics to which the Dalai Lama often liked to turn.

A few weeks before the event, I learned along with the other Maitripa College students that the symposium was to be titled "Life after Life" and focus on the topic of human rebirth. Upon reading the email sharing this news, my shoulders tightened and I immediately entered a defensive and skeptical mindset. This was an area that Jim Blumenthal and I had discussed on frequent occasion around Maitripa College and during some Tarn Tip lunches in Corvallis, and it was an area where my view seemed to be entirely and irreparably at odds with the consensus Buddhist view. My mind had concluded that human rebirth was an historical artifact of Buddhism that should be discarded and not thought about. It was a topic I simply wanted to avoid. I was now more discouraged than excited for the event, but I knew I would, of course, still attend.

I arrived at Maitripa College at 5:30 am on the day of the symposium, three and a half hours before the Dalai Lama's scheduled arrival time. We were told this was necessary for security; the police cordoned off the city blocks immediately surrounding Maitripa College in the Ladd Circle neighborhood of Portland. I had stayed the night before with lifelong friends in Portland, Mike Schurke and Sona Pai and their family. After parking many blocks away and walking to the college, I saw a large community of well-wishers already gathered at the entrance, awaiting the Dalai Lama's arrival. I navigated through the crowd and into the Maitripa College building, and then joined the other students and invited guests in the *Jokhang*. In addition to my classmates, the audience included folks appearing as Portland hipsters, older white Buddhist appreciators and practitioners, as well as Tibetan monks and nuns. Yangsi Rinpoche and Jim Blumenthal were both busy welcoming honored guests and guiding people here and there. I found my assigned spot in the *Jokhang*, pulled out my recently purchased Zen bench, and like the other students arranged in the front rows of the *Jokhang*, clumsily tried to orient myself on the unfamiliar wooden meditation device. Upon getting situated in something approaching bodily balance on the bench, physical pain shot through my knees and legs; I stressed out about the prospect of this *duhkha* continuing for many hours ahead. But, after fifteen minutes, the pain either subsided or became more tolerable, and it became less of a preoccupation to my mind. After arriving and getting settled, audience members were instructed to maintain quiet and meditate. Later, the group was led in Tibetan chanting, cycling through a stanza of the *Diamond Sutra* that is incorporated into preparatory chanting for upcoming teachings:[48]

> KAR MA RAB RIB MAR ME DANG
> A star, a visual aberration, a flame of a lamp,
> GYU MA ZIL PA CHHU BUR DANG
> An illusion, a drop of dew, or a bubble,
> MI LAM LOG DANG TRIN TA BUR
> A dream, a flash of lighting, a cloud . . .
> DU JA CHHO NAM DI TAR TA
> See all conditioned things as such!

Two honorary guests then arrived, the Buddhist scholar José Cabezon from the University of California, Santa Barbara, and Eben Alexander, a neurosurgeon and author of *Proof of Heaven*,[49] a popular book that recounted the doctor's near-death experience where he claims to have experienced an afterlife. Then, a contingent of Tibetan monks and nuns entered the *Jokhang* and were seated by event organizers, followed by the Dalai Lama and his translator. The Dalai Lama was upbeat, as always, doling out hugs and hellos to the Tibetan monastics as he walked in. It had the unexpected feel of a high school reunion. A very small, elderly monk teared up after the Dalai Lama's entrance and embrace; I was later told that this was one of the Dalai Lama's childhood monastic classmates in Tibet, and someone that the Dalai Lama hadn't seen since they both fled across the Himalayas from the Chinese invasion in 1959.

The Dalai Lama then took his seat, front and center with his translator nearby as usual, and settled in for introductory remarks. José Cabezon was seated to their right, and Eben Alexander was seated to their left. The game plan for the symposium was outlined by an event organizer: the two honorary participants would each provide short discussions of their respective experiences or study of human rebirth, followed by a longer lecture by the Dalai Lama on the topic. Dr. Alexander went first and highlighted his personal experiences in a coma caused by acute meningitis. He described an experience of grand awareness during this time, attributed to his mind being liberated from his brain, that transformed the neurosurgeon's view of past and future lives and the potential of consciousness. Professor Cabezon went next and provided an academic tour through historical attempts to prove human rebirth in ancient India. He highlighted the work of Dharmakirti, a highly influential sixth-century CE Buddhist scholar who worked at Nalanda University in India. Dharmakirti, in debating against those posing more materialist views in opposition to human rebirth, asserted that consciousness was not of the same "nature" as matter and exists in a separate continuity. Dharmakirti argued that, because no *dharmas* can truly be created or destroyed as per *pratityasamutpada*, the stream of consciousness must go somewhere after biological death.[50] Cabezon also discussed the rarity of rebirth accounts in ancient India due to an expectation of sacred privacy surrounding such phenomena.

The Dalai Lama talked next. He started by lauding the reliability of scientific research but also noted that there are some phenomena, rebirth among them, that are beyond science's capacity to investigate effectively. He outlined three categories of phenomena that underpin Tibetan Buddhist conceptualizations of reality: evident phenomena, subtle phenomena, and extremely subtle phenomena.[51] The Dalai Lama expounded that science generally deals with what is evident, but on the basis of inference also touches on subtle phenomena, an example being Darwin's theory of evolution by natural selection. The third class, extremely subtle phenomena, with regard to the mind, is only accessible in the context of one's subjective experience. It then becomes a matter of personal experience and trust in teachers, with reliability dependent upon on the credibility of the informant. The Dalai Lama pointed out that some phenomenon, such as rebirth, being beyond science's ability to investigate is not the same as science demonstrating it not to be so. During the symposium he said:

> My own observation is that science is beginning to explore the importance of our various mental states. I've been having serious conversations with scientists for more than thirty years and these investigations are still in the initial stages. Trying to formulate a theory of consciousness, which is a very subtle phenomenon, only on the basis of a material organ, the brain, is problematic.

At the conclusion of the event, the Dalai Lama rose from his chair and began circulating around the group of students seated on Zen benches in the *Jokhang*, asking them to rise as a group for departing hugs. I rose from my bench and discovered that my legs had become completely numb; they almost buckled underneath me. A hand grabbed my arm to help steady me; my head turned toward the Dalai Lama's smiling face. He then squeezed my arm with both hands, rubbed the back of my bald head, and then moved on to greet other students. On his way out the door, the Dalai Lama paused by the three hundred volumes of Buddhist literature in the *Jokhang*, noting that these Tibetan tomes constitute written records of Buddhist science.

* * *

The Dalai Lama's explanation changed my view of human rebirth; it is now a topic that I like to interrogate rather than avoid. There continues to be a void of reliable scientific evidence in support of this existential underpinning of much of Buddhist philosophy and culture, at least in my mind and I imagine also in the minds of most of my Western scientific colleagues. However, the Dalai Lama's framing of human rebirth as an "extremely subtle phenomenon" where evidence is only available through direct subjective experience during advanced meditative equipoise has transitioned my stance on rebirth away from "almost certainly not" to "I haven't done the right experiments, so I don't really know."

An analogy is provided by our understanding of the mitochondria that exist inside the cells of animals, plants, and many other life forms. These cellular substructures, or organelles, are widely accepted entities that perform central metabolic functions inside the cells in which they reside. The existence of mitochondria, however, required the invention of microscopes providing scientists the necessary lens of magnification to find evidence for their existence.[52] Without microscopes or similar devices, scientists would have no means by which to detect mitochondria or any other microscopic features of life. Similarly, the Dalai Lama and other Buddhists claim that advanced meditative lenses, which are entirely subjective in nature, are required for one to detect evidence for human rebirth. Although I am a biologist, and one who has extensively studied mitochondria and the DNA residing inside these organelles at that, I have never personally observed mitochondria through a microscope. Instead, I rely on the multiple observations and experiments done by others in the past and ongoing today, scientists in whom I place trust, for my operational assumption that mitochondria are real. Should similar trust extend to accounts from Buddhist masters of meditation? I'm still trying to figure it out.

6

Sciences

In addition to sharing historical and ongoing interactions between Buddhist and biological thinkers, a second major objective of the present work is to directly evaluate core concepts from Buddhist teachings, using a formal scientific approach. As detailed in the previous chapters, many scholars from both the side of Buddhism and the side of science have identified striking harmonies between Buddhist and scientific outlooks on the nature of reality and the universe. Positive tones of unexpected synergy (e.g., Barash's Buddhism and science POMA) often dominate the Buddhism and science scene, though caveats and qualifiers always accompany the enthusiasm. Authors coming from the side of Western science generally operate in this dialogue with outright dismissal of the Buddhist concept of human rebirth,[1] and the various supernatural beings (e.g., deities, nagas) found in *sutras* and other parts of Buddhist culture. Authors coming from the side of Buddhism often comment on science's positive contributions to human health and well-being. Some, such as the Dalai Lama, also recognize value in science's capacity to discern basic truths. However, Buddhist writers also often characterize science as a tradition that is "materialistic,"[2] incomplete and still in immature stages of infancy, and deficient in ethical and moral rooting. In addition to vaccines and antibiotics, science also gave us automatic rifles and atomic bombs.

Although the manifold accounts of harmonies along the Buddhism-science interface offer promise and reasons for further pursuit of insights along this intersection, a pointed and direct analysis of core Buddhist tenets from scientific lenses remains absent, to the best of my knowledge. Such an investigation is essential for many science minds to find true value in Buddhist principles and motivate them to become more active participants in advancing this scholarly frontier. This chapter and the two that follow will explicitly evaluate core teachings of the Buddha from a scientific standpoint, using a hypothesis-guided approach. Here at the onset of this adventure, it is important to present and consider the specific approach by which the inquiry will be carried out and the historical underpinnings from which it arose. There are multiple meanings and interpretations of "science," and many varieties of scientific methods. Further, given the topic at hand, it is of special importance to ensure that the investigation does not fall into pseudoscientific pitfalls.

The Methods of Science

Similar to the varying identities of the truth (e.g., conventional truth and ultimate truth) held by Nagarjuna and other Madhyamaka philosophers, there exist multiple conceptual identities for science. In the early education of many young people in the West, however, the scientific process is often presented as a single, linear, invariant, and universal "scientific method," composed of four or five basic steps. In the philosophy of science literature, this is often referred to as *the* scientific method.[3] The process of this method is typically taught as starting from some observation that then leads to a question. This initial step progresses to the formulation of a hypothesis aimed at evaluating or explaining the observed phenomenon, or the question derived from it. Hypotheses are then examined, often by experiments designed by a scientist. After experimental results are analyzed, conclusions are drawn. This simplified model of a single and universal methodology for the practice of science, the legend of *the* scientific method, is prevalent throughout textbooks and other learning materials across all levels of Western education systems. It is a prominent shaper of how people conceptualize science. One need not look far, however, to identify examples of activities that many would usually describe under the banner of "science" but deviate from the simplified formula of *the* scientific method.

* * *

J. Craig Venter is a science paradigm disrupter. This headline-grabbing scientist's contributions have ranged from providing the first draft of a human genome sequence to creating synthetic life. In his early days, Venter didn't take education seriously; he instead preferred to play in the water, by boating and surfing.[4] After time in the Vietnam War where he worked in the intensive care units of field hospitals, Venter returned to the United States to pursue higher education and focused on biomedical technologies. He started at a small California community college, College of San Mateo, and then transferred to the University of California, San Diego, where he eventually earned BS and PhD degrees.

Venter famously led a private company, Celera Corporation, in the race to provide the first complete draft of a human genome during the late 1990s. Venter developed an innovative new "shotgun" DNA sequencing strategy that offered a much faster route to assembling a complete genome sequence,[5] as compared to the slower and more methodical approaches of ongoing efforts in government and university research labs. At first, Venter's shotgun strategy was thought likely to succeed on the smaller genomes of bacteria and flies, but predicted to fail when applied to the enormous and complex human genome. Unperturbed, Celera Corp. launched its private effort to sequence the human genome in direct competition with the long-standing government-funded efforts, resulting in the latter group of scientists ramping up the pace of their work in turn. Along the way, Celera's stock prices rose, fueled by selling early access to human genome information and the promise of future profits deriving from this unprecedented genetic resource. On June 26, 2000, then-President Bill Clinton announced the completion of the first virtually (~99 percent) complete draft of the human genome. As the president arrived to deliver this historic speech, he was flanked by Craig Venter on his right and Francis Collins, the leader of the publicly funded genome-sequencing effort, on his left. A diplomatic tie was declared in the public and private efforts to sequence the human genome.[6]

Shortly after his time in the human genome media spotlight, Venter was sacked by Celera. The company's stock prices took a nosedive after the human genome information became freely available.[7] The hypercompetitive scientist had now drawn disdain from both the public and private scientific sectors. Venter escaped to St. Bart's in the comfortable waters of the Caribbean on his private yacht, *Sorcerer II*, to lick his wounds and dream up new grandiose scientific plans. One of these new bold ambitions concocted by Venter was to sequence every gene on the planet. He launched a pilot project whereby he sequenced "ocean water" DNA using the same basic shotgun strategy previously applied to the human genome. Venter targeted the genetic material residing inside the diverse and largely unknown marine microbes living in the Caribbean. This initial effort resulted in the discovery of more than a million previously unknown genes residing inside the diverse community of bacteria and other microscopic life forms inhabiting ocean waters.[8] Thousands of new microscopic marine species were discovered. After finding success with this pilot project, in 2004 Venter gathered a crew for

Sorcerer II and began 32,000 nautical mile and two-year adventure around the world. The bald, white scientist grew a beard and circumnavigated the globe in his Speedos (or sometimes buck naked), stopping along the way to sequence buckets of random DNA fragments from the microscopic life forms living in the ocean water. Subsequent voyages built upon this early expedition and further expanded our knowledge of previously hidden oceanic genetic diversity. This tour de force was a game changer in science; it opened up biologists' eyes to the vast previously unknown diversity and evolutionary potential of marine microbial life, and it demonstrated the feasibility of environmental investigation using approaches almost exclusively reliant on DNA. Venter credited his ego, his enormous sense of self, as a major necessary force in achieving his scientific ambitions.[9]

Despite the undeniable impact that Venter's luxury scientific yacht voyage had on modern twenty-first-century biology, it was completely devoid of central, guiding hypotheses. Further, the human genome project efforts, both public and private, were motivated to generate an unprecedented genetic resource, but not to evaluate a core scientific hypothesis. These and similar exploratory endeavors are often described as "discovery science"— efforts that are primarily focused on uncovering new knowledge without an explicit overarching hypothesis to guide the effort. Do hypothesis voids in Venter's oceanic adventures, and others that deviate from the core scientific method paradigms portrayed in textbooks, implicate such activities as nonscientific in some way? If still considered scientific, does the absence of hypotheses relegate them to a lower tier on the scientific hierarchy? What exactly are the features that justify placing particular activities under the umbrella of "science"? Questions such as these motivated many centuries of Western philosophical discourse focused on the nature of science and its methods.

* * *

When asked, both practitioners of science and philosophers of science can often agree on a set of activities that are characteristic of science. For example, undertakings described as "scientific" generally involve systematic observation and/or experimentation. Observation and data collection steps are usually followed by some kind of interpretation and intellectual reasoning. *The* scientific method commonly presented in textbooks and other science educational materials encompasses these two general features, though in a more explicit and specific fashion that also involves a hypothesis.

The core components of the scientific method are generally considered as separate from other external features that are often deemed necessary features of "good" science. These latter features are often referred to as scientific "meta-methodology" in the philosophy of science literature. Examples of meta-methods typically valued in science include objectivity, replication, reproducibility, and simplicity. Further, the specific experimental procedures and protocols (for example, sequencing DNA from buckets of ocean water) are considered as distinct from the generalized steps of the scientific method.[10]

The scientific method and meta-methods, highly valued by practicing scientists, resulted from many centuries of thought development in the Western world. Approximately one century before the Greek thinker Pyrrho encountered wandering naked wise men during his India excursion in ~334 CE, another Greek philosopher, Plato, was laying the foundations for science and broader Western thought and culture. One of Plato's key developments was to articulate a distinction between what was *visible* versus what was *intelligible*.[11] The latter led to Plato's famous Forms; these Forms constituted abstract, conceptual, and idealized mental constructs, which were the valid objects of knowledge. Plato reasoned that observations of the natural world held the potential to be deceptive and constituted only shadows of reality,[12] as expressed in the cave allegory of Plato's *Republic*. The differentiation between observation and reasoning established by Plato set the stage for further development by his successors in classical Greek thought.

Aristotle, student of Plato, is widely credited with providing the world's first systematic treatise on the nature of scientific inquiry through his *Prior Analytics* and *Posterior Analytics*.[13] The former work focused on the aims of inquiry; the latter considered the methods. Unlike his teacher Plato, Aristotle considered the true Forms to reside in the natural world. Aristotle recognized that the aims of a given inquiry were important determinants of the particular methods that should be deployed. He promoted empirical approaches to investigation, valuing methodologies involving careful and passive observation of the natural world. Three basic steps defined Aristotle's approach: discovery, ordering, and displaying. The process began with the investigator neutrally observing some phenomenon in the natural world, followed by intellectual reasoning and organization of observations using some system of logic, and the final arrangement or display of findings in a fashion that is clear and interpretable. Aristotle organized his broader system of logic and reasoning in the *Organon*, a compendium that included methods such as induction, analogy, and prediction (as well as his writings on fallacious forms of reasoning). This title would serve as a foundation upon which subsequent Western works on science and reasoning would build, such as Francis Bacon's *Novum Organon* written twenty centuries later in 1620.

Aristotle developed a general theory of cause and effect that underpinned his analytical methodologies, as well as subsequent Western thought on this topic across the centuries.[14] In *Posterior Analytics*, Aristotle proposed that understanding causes was of central necessity in establishing valid knowledge on a particular subject of study. In *Physics* and *Metaphysics*, Aristotle formulated a set of four types of causes to be applied to any phenomenon or object of study requiring explanation. He commonly deployed objects of art and the human body as examples. Aristotle's four categories of cause were as follows:

- Material Cause: "that out of which" a particular object is made. Examples include the bronze of a statue and the clay of a pot.
- Formal Cause: "the form" or "the account of what is to be." Examples include the shape of a statue and the form of an adult human being.
- Efficient Cause: "the primary source of the change." Examples include the artisan who creates a bronze statue, the societal tradition of statue making in ancient Greece, and the mother of a child.
- Final Cause: "the end goal, the sake for which a given thing is done." Examples include the honoring of a Greek general whose physical form is represented in the statue, and sitting in meditation to improve one's health.

For Aristotle, these four causes were not equal. In *Parts of Animals*, a treatise focused on the investigation of animal life and written after *Posterior Analytics*, Aristotle argued that the Final Cause is of higher explanatory priority as compared to the Efficient Cause.[15] He reasoned that it was impossible to explain the sources of change without reference to the final form at the end of the process. Aristotle deployed analogy as a method to make his case, comparing artistic production to the process of becoming (i.e., birth) for animal life. There is no other way to explain how a sculpture is created without in some way referencing the final result of the process, the sculpture itself. Aristotle claimed this was also true for the production of animal life. In *Parts of Animals*, Aristotle claimed, "Generation is for the sake of substance, not substance for the sake of generation." As a consequence, for Aristotle the proper and primary way to explain the cause of the generation of an animal, or the parts of an animal, was by reference to the end product (i.e., the animal form) at the end of the process. The privilege and primacy afforded the final form in Aristotle's cause-and-effect framework contrasts sharply with the *anatman* (non-self) teaching of the Buddha. This distinction was evident in the dialogue between the Buddhist monk Nagasena with the Greek King Menander, discussed earlier in Chapter 2.

Aristotle established an inductionist baseline upon which subsequent scientific theory and development would build. In general, induction describes a process whereby particular cases or objects of study are observed, followed by the formulation of general concepts based upon the previously observed specific cases. In Aristotle's three-step approach to inquiry, an initial set of induction events served as discovery foundations upon which the two subsequent steps, ordering and displaying, were built. Inductive reasoning, however, would eventually fall from favor and give way to deductive reasoning approaches and the dawn of the hypothesis as drivers of scientific investigation.

* * *

In the *Organon*, Aristotle subdivided the process of reasoning into two primary forms; this rough division persisted throughout the subsequent centuries of scientific thought, into modern times.[16] These two basic forms eventually evolved into what is known today as induction and deduction. The fundamental distinction between these two approaches centered in the pathways of inquiry: inductionism proceeded away from what was observed, toward general principles; deductionism started from the general principles and flowed toward specific examples or instances of those principles. Giants of the sixteenth- and seventeenth-century European Scientific Revolution, such as the Englishmen Francis Bacon and Isaac Newton, built upon Aristotle's inductionism toward theories and approaches that more closely resemble modern scientific methods. In *Novum Organon*, Bacon advocated a careful and cautious inductive approach, grounded in methodical observation and data collection, coupled with the correction of senses and methods of data collection along the way to avoid various types of errors. With Bacon's approach, the scientific community could carefully and gradually climb toward reliable and generalizable claims. Bacon's approach was widely criticized, however, for being far too impractical and blind to the working realities of practicing scientists. Isaac Newton developed an inductive approach that was grounded in a mathematics-rooted philosophy combined with practical experimental science. Newton based his system on quantitative observations, not previously reasoned first principles. To this end, Newton famously stated "*hypotheses non fingo*" (commonly translated as "I frame no hypotheses").

Paradigm shifts in science continued after the Scientific Revolution, including Darwin's theory of evolution by natural selection that rocked the world of biology in the 1800s, and the quantum revolution in physics of the 1900s. Alongside these fundamental breakthroughs in our conceptual understanding of how life and the universe work, inductive scientific approaches started to give way to deductive methods. Consensus started to build that purely inductive approaches, along the lines of those advocated by Bacon and Newton, were fundamentally unsound. The conclusion that there was, in fact, no such thing as "pure" observation fueled the growing consensus against inductionism. Some kind of theory or background knowledge was always a preexisting condition for thinkers making observations. Carl Gustav Hempel, a twentieth-century CE philosopher of science from Germany, developed the "raven paradox" to illustrate one of the specific problems with inductive reasoning along these lines, by linking it to the process of intuition. Hempel wrote that if an observation confirmed a given generalization, it simultaneously confirmed all other generalizations that were logical equivalents.[17] The statement "All ravens are black" was the same as saying "All non-black objects are non-ravens." Given this, the observations of a black raven, a white yacht, and a green tea cup would all equally contribute to supporting the generalization that ravens are black. Using this paradox, Hempel argued that a thinker's intuition is grounded in inescapable preexisting background knowledge on the nature of ravens and non-ravens. This intellectual solvent leads one to ascribe more privilege to evidence in support of ravens being black, as compared to evidence of non-black objects or phenomena (or in Buddhist terms, *dharmas*) being non-ravens.

The downfall of inductionism coincided with the advancement of deductionist approaches in the nineteenth and twentieth centuries. The generalized "hypothetico-deductive" approach, whereby a statement expressing some hypothesis was evaluated based on whether or not its expected consequences were true, served as a standard starting reference point for these more modern deductive approaches to inquiry.[18] This ultimately led to the development of Karl Popper's falsification approach to scientific investigation.[19]

Popper, an academic born in Austria who did most of his scholarly work in England, is widely regarded as one of the greatest twentieth-century philosophers of science. The Popperian method purported that, regardless of how much evidence was gathered in support of a given hypothesis, a scientist never achieved absolute certainty that the hypothesis was "true" or "proven," without committing the fallacy of affirming the consequent. This mistake of logic, also sometimes referred to as "converse error" involves taking a true conditional statement and invalidly inferring its converse to be true. For example, the true statement "If the lamp is broken, then the room is dark" might lead a thinker to erroneously conclude the inverse, "The room is dark, so the lamp is broken." A broken lamp offers one possible explanation for the dark room, but there are many other valid potential explanations, such as the lamp is turned off, the lamp is not plugged in, or the electrical outlet is broken. Popper supported corroboration as a measure for how well a hypothesis has survived previous testing; as a hypothesis underwent more and more evaluation through more and more scientific approaches and valid analytical perspectives, and continued to not be falsified through this ongoing gauntlet of critical evaluation, the more support the hypothesis gained. This basic approach underpins most applications of the scientific method used by practicing scientists today. Many modern scientists say the name "Popper" with reverent tones of profound respect, and assign ultimate authority to his deductive, hypothesis-based falsification framework.

The Methods of Pseudoscience

In addition to providing a sound scientific method, Popper was motivated to clearly demarcate science from pseudoscience.[20] From Popper's view, the central validity of science derived from the act of subjecting hypotheses to rigorous tests which offered a high probability of failing. The goal of science and scientists was not, in the Popperian approach, to verify a given hypothesis or theory, but rather to proactively and continuously seek avenues to reject it. This contrasted with the aims of pseudoscience, whose perpetrators were often accused of freely and frequently adding ad hoc modifications or other adjustments to buffer and ensure continuing support for their favored hypotheses. Popper was particularly concerned about various schools of psychoanalysis and Marxist theories; he charged that they made statements that were not falsifiable and were thus unscientific.

The term "pseudoscience" brings with it the most damning, derogatory label that can possibly be attached to any knowledge seeking endeavor. This contrasts with the legitimacy and respect that typically accompanies the label "science." Significant social repercussions can accompany efforts to discern science from pseudoscience, exemplified by the 2005 *Kitzmiller v. Dover Area School District* legal case, a US federal court trial testing a public-school policy requiring the instruction of intelligent design. The verdict in this landmark case ruled against intelligent design, with major influence arising from two philosophers of science, Barbara Forrest and Robert Pennock, who provided expert testimony that intelligent design was in fact a pseudoscience, not science.[21] Given the fundamental and widespread importance of distinguishing between science and pseudoscience, philosophers of science have invested substantial effort into defining universal demarcators between the two. Although there has been strong consensus on assigning the pseudoscience label to specific cases (e.g., astrology, alchemy, eugenics), the broader goal of defining basic criteria that can universally be applied to distinguish science from pseudoscience has proven difficult. One major challenge has been framing the science-pseudoscience distinction in terms of the broader boundaries between science and all other non-science activities.[22] For example, art is generally not considered a science but is also generally not considered a pseudoscience.

One approach to categorizing the interrelationships of science, non-science, and pseudoscience considers which particular scholarly subjects and activities belong to a broader "Community of Knowledge Disciplines."[23] Those subjects belonging to this community constitute disciplines commonly encountered at institutions of higher education, including those that fall under the broad category of the "humanities" (e.g., art, philosophy, Buddhist Studies) and those that represent formal traditional "sciences" (e.g., physics, chemistry, biology). All of these disciplines are generally held as trustworthy contributors to knowledge and understanding.[24] Pseudoscience, however, is not a member of this community. Thus, under the Community of Knowledge Disciplines model (see Figure 6.1), the humanities are allied with the sciences in this community of reliable sources of scholarly knowledge but are also fall into the category of "non-science" along with pseudoscience.

Some efforts to define universal science-pseudoscience separation criteria have focused on portraying pseudoscience as a science poser. Following this approach, two defining features of pseudoscience were proposed: (1) the activities are nonscientific, and (2) their proponents try to create the impression

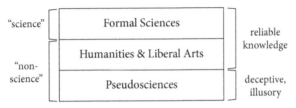

Figure 6.1 Community of Knowledge Disciplines. Adapted from Mahner (2007).

that the activities are scientific.[25] This simple separation strategy offered a wide umbrella that aptly fit all of the well-accepted examples, but it could also be considered too broad because there were activities that fit both criteria (e.g., scientific fraud and intentional data falsification) but were not generally considered pseudoscience. Further philosophical work aimed to establish demarcators that included all the well-established pseudosciences while simultaneously ensuring that nonpseudoscientific activities were excluded.

Popper presented his hypothesis falsification framework as a way to draw the line between statements belonging to the empirical sciences and "all other statements—whether they are of a religious or of a metaphysical character, or simply pseudoscientific."[26] However, Popper's falsification-based criterion has been criticized for both excluding some genuinely scientific activities and for lending scientific status to some pseudoscientific activities.[27] In fact, Popper once advocated that biological evolution by natural selection is not a proper, testable scientific theory but instead a "metaphysical research program." This claim elicited the ire of the evolutionary community and drew many responses defending the valid scientific nature of the theory of natural selection. Defenders pointed out that the theory has given rise to numerous derivative hypotheses and predictions that have survived multiple varied forms of testing in the lab and in nature.[28] Popper later retracted this controversial view of natural selection in a 1977 lecture at Darwin College.

The struggling quest to define clear, universal demarcators between science and pseudoscience led to the development of alternative approaches whereby multiple criteria were used to delineate the distinction. Many lists of pseudoscience criteria were generated by philosophy of science thinkers during the latter half of the twentieth century and early twenty-first century,[29] and often included many common features. In 1993, A. A. Derkson identified the pseudoscientist, the practitioner of pseudoscience, as the key problematic player and created a list of "seven sins of pseudoscience":[30]

1. *Dearth of decent evidence*: pseudoscientists often pretend to produce reliable knowledge via trustworthy methods, when in reality the evidence presented is unreliable.

2. *Unfounded immunizations*: pseudoscientists often immunize their claims by tinkering with empirical data found to be inconsistent with the idea, making the observations more consistent with the original claim.

3. *Ur-temptation of spectacular coincidences*: pseudoscientists often succumb to the error of uncritically assigning deeper meaning to instances of coincidence.

4. *Magic methods*: pseudoscientists often forego the accepted methodologies of the formal sciences and instead create their own novel methods which are specifically molded to generate the data they need to support their claims.

5. *Insight of the initiate*: pseudoscientists often claim to bear a special insight and training, and that only those individuals with this type of preparation are capable of seeing the truth.

6. *The all-explaining theory*: pseudoscientists often wield all-encompassing, universal theories that they claim explain everything; these theories are usually unfalsifiable.

7. *Uncritical and excessive pretension*: pseudoscientists often ascribe excessive reliability to their claims despite contrary evidence (or lack thereof); they also assign undeserved special importance to their pseudoscientific theories.

* * *

The important task of distinguishing science from pseudoscience is further complicated by the fact that some pseudosciences give rise to valid scientific disciplines, and some sciences give rise to deceptive and damaging pseudosciences. The development of the formal science of chemistry from a previously existing pseudoscience, alchemy, offers an example of the former. For the latter, one of the most embarrassing and deeply disturbing developments of the early twentieth century was the rise of eugenics, a pseudoscience relying on bad data and faulty assumptions that aimed to improve the genetic quality of the human population through selective reproduction. Eugenics, deriving from the intertwined biological subdisciplines of evolution and genetics, became a regular component of university curriculum and research across the Western world in the early 1900s. Eugenic "science" was also used as a source of justification cited by purveyors of horrors such as state-enforced sterilization policies targeting the "feeble-minded" and the "sexually deviant" in the United States and other countries, and the genocide of the Nazi Holocaust.[31] These examples illustrate the reality that pseudoscience can emerge and take root right in the hallowed halls of academic science. Scientists, like anyone else, are susceptible to the allure of pseudoscience and must proceed with caution and mindful awareness, especially when paddling into new waters.

Buddhist Hypotheses

A central objective of the present work is to evaluate foundational teachings of the Buddha, subsequently supported and expounded upon by other Buddhist thinkers across the last twenty-five centuries, using a formal scientific approach. Five specific teachings of the Buddha have been selected for preliminary consideration of evaluation, based on their centrality in most forms of Buddhism and expression of fundamental Buddhist doctrines. These include Buddhism's "three marks of existence": *duhkha* (suffering, unsatisfactoriness), *anitya* (impermanence), and *anatman* (non-self). In addition, *pratityasamutpada* (dependent arising), the mutual cause-and-effect framework that underpins Buddhist thinking, is added to the list. Last, claims regarding recurring human rebirth (*punarbharva*, Skt.; *punabbhava* in Pali), a phenomenon that is central to most but not all,[32] Buddhist traditions, will be initially considered for examination. All five of these concepts are hallmark, defining features of virtually every branch of Theravada and Mahayana Buddhism. Only three of these teachings, however, will survive consideration and move on to formal analysis in the subsequent chapters.

A Popperian-style, hypothetico-deductive approach will be applied to evaluate those teachings of the Buddha that move on to full analysis. A hypothesis development process will ensure that the statements to be examined are well suited to the specific approach (focused on DNA) to be employed here, and ultimately determine which teachings move on for analysis and which ones are set aside. As a first step, each of the five Buddhist concepts will be expressed as hypothesis statements. General candidate hypotheses will first be formulated; the main statements will be expressed in English, with relevant Sanskrit terms provided in parentheses where relevant. Each general hypothesis will then be transformed into a corresponding specific hypothesis, making reference to the subject of study chosen for this work: DNA. After the initial candidate general and specific hypotheses are stated, their falsifiability and suitability, relative to the specific investigative approach deployed here, will be evaluated. Then, the hypotheses will be reorganized and assigned specific alphanumeric designators that will be used in downstream analyses. Once defined, evaluated, and reorganized in this chapter, hypotheses evaluation will proceed in subsequent chapters. At the end of the analysis, the overall approach and methodology employed here will be re-examined to ensure that pseudoscientific sins were not committed.

* * *

The five general Buddhist hypotheses to be initially considered as candidates for evaluation are as follows:

- All phenomena (*dharmas*) experience suffering (*duhkha*), as described by the Buddha and other Buddhist teachings.
- All phenomena (*dharmas*) are impermanent (*anitya*), as described by the Buddha and other Buddhist teachings.
- All phenomena (*dharmas*) lack a self-nature (*anatman*), as described by the Buddha and other Buddhist teachings.
- All phenomena (*dharmas*) exist in a mutual cause-and-effect framework of dependent arising (*pratityasamutpada*), as described by the Buddha and other Buddhist teachings.
- All phenomena (*dharmas*) undergo rebirth (*punarbhava*), as described by the Buddha and other Buddhist teachings.

The five specific Buddhist hypotheses, all focusing on DNA as the subject for subsequent analysis, to be considered as candidates for evaluation are as follows:

- DNA experiences suffering (*duhkha*), as described by the Buddha and other Buddhist teachings.
- DNA is impermanent (*anitya*), as described by the Buddha and other Buddhist teachings.

- DNA lacks self-nature (*anatman*), as described by the Buddha and other Buddhist teachings.
- DNA exists in a mutual cause-and-effect framework of dependent arising (*pratityasamutpada*), as described by the Buddha and other Buddhist teachings.
- DNA undergoes rebirth (*punarbhava*), as described by the Buddha and other Buddhist teachings.

Upon initial inspection of the candidate hypotheses stated earlier, two of the specific statements immediately stand out as problematic: the first one and the last one. Most Buddhist conceptualizations of *duhkha* and rebirth specifically focus on human beings as the subject. *Duhkha* relies heavily on the lived human experience, as expressed through the five *skandhas* (aggregates) and Four Noble Truths framework. Though DNA has the potential to play roles in suffering experienced by human beings, it is difficult to conceptualize the possibility that these hereditary molecules (or other biomolecules such as proteins, lipids, and sugars) somehow experience "suffering" in a sense similar to that of conscious human beings. The proposed process of human rebirth (*punarbhava*) in Buddhism relies heavily on Buddhist concepts of the nature of consciousness, which has virtually nothing to do with DNA (at least in a direct sense). Thus, it is concluded that the first and last candidate specific hypothesis statements listed earlier are not suitable for examination in the DNA-centric approach of this work and hereafter will be abandoned. The decision to discard these two specific hypotheses constitutes a necessary limitation that accompanies the chosen methodology to be deployed in this study. Further, the decision to limit the analysis to only those teachings with relevance to DNA serves as a preventative measure against the sixth of Derkson's seven pseudoscientific sins: *the all-explaining theory*.

The remaining three specific hypotheses—focusing on *anitya*, *anatman*, and *pratityasamutpada*—remain valid candidates for evaluation using DNA as a relevant test subject. Any permanent, unchanging aspects or features of DNA identified will falsify the specific hypothesis focused on *anitya*. Any inherent, intrinsic features of DNA that suggest some aspect of existence outside the context of interactions with other biomolecules will falsify the specific hypothesis focused on *anatman*. Any experimental observations of DNA that suggest "single causes," true "first causes," or existence in a purely linear rather than mutual cause-and-effect framework will falsify the specific hypothesis focused on *pratityasamutpada*. Thus, it is concluded that these three specific hypotheses remain suitable, falsifiable statements that are well suited to the DNA-centric investigation to follow.

Given the hypothesis development efforts of the preceding text, the surviving hypothesis statements will now be reordered and named to facilitate evaluation efforts in future chapters. The three surviving general hypotheses will be listed and numbered first, followed by the three corresponding remaining specific hypotheses to be evaluated in-depth in the chapters that follow:

Final General Hypotheses

H_G1: All phenomena (*dharmas*) are impermanent (*anitya*), as described by the Buddha and other Buddhist teachings.

H_G2: All phenomena (*dharmas*) lack a "self" nature (*anatman*), as described by the Buddha and other Buddhist teachings.

H_G3: All phenomena (*dharmas*) exist in a mutual cause-and-effect framework of dependent arising (*pratityasamputpada*), as described by the Buddha and other Buddhist teachings.

Final Specific Hypotheses

H_S1: DNA is impermanent (*anitya*), as described by the Buddha and other Buddhist teachings.

H_S2: DNA lacks a "self" nature (*anatman*), as described by the Buddha and other Buddhist teachings.

H_S3: DNA exists in a mutual cause-and-effect framework of dependent arising (*pratityasamputpada*), as described by the Buddha and other Buddhist teachings.

7

Molecules

Three specific Buddhist hypotheses were defined in the previous chapter, and all focused on DNA as the target of analysis. The choice to limit the investigation to DNA resulted in a trade-off. Two key Buddhist doctrines, *duhkha* (suffering) and *punarbhava* (rebirth), depended deeply upon human consciousness and have little to nothing to do with DNA, at least in a direct sense. Thus, they were necessarily jettisoned from the center of this DNA-centric analysis. On the upside of the trade-off, three strong and falsifiable hypotheses on the nature of DNA remained, focusing on the Buddha's core propositions of *anitya* (impermanence, H_S1), *anatman* (non-self, H_S2), and *pratityasamutpada* (dependent arising, H_S3).

DNA offers many compelling advantages as a focal study subject. This molecule is regarded by many to be the most important, "driving" factor in biological processes. When asked to name one single molecule associated with life, "DNA" is certain to be the most common answer. The story of DNA is well known, offering classic experiments and quintessential features that will be deployed here in efforts to falsify the three specific Buddhist hypotheses. There exists a massive compendium of peer-reviewed scientific research articles published about DNA; a 2021 search of the PubMed database of the US National Library of Medicine revealed 1,719,487 entries when "DNA" was used as a query search term.[1] Biologists thus know a lot about DNA and how it interactively functions with other biomolecules. Further, DNA has connections to personal identity (e.g., forensic applications and direct-to-consumer commercial genetic testing kits), and it is thus particularly well suited to serve as a vehicle in efforts to falsify the focal hypotheses.

The three specific hypotheses defined in the previous chapter lead to specific predictions about the nature of DNA and paths to falsification. H_S1 predicts that all features of DNA are impermanent in nature. Any permanent, unchanging aspects of DNA detected during the investigation would falsify the first specific hypothesis. H_S2 predicts that DNA lacks inherent features that would be characteristic of a "self" nature for the molecule. Any truly intrinsic, invariant aspects of the DNA detected would falsify the second hypothesis. H_s3 predicts that DNA exists and functions in a mutual cause-and-effect framework, consistent with that outlined by the Buddha and other Buddhist teachers.[2] Three specific paths to falsification are available for H_s3: the detection of a true "first cause" of DNA, a "single cause" associated with some aspect of DNA, or evidence of linear cause or effect involving DNA that is inconsistent with the "mutual" cause-and-effect aspect of *pratityasamutpada*.

The road to evaluating the three specific Buddhist hypotheses will tour through classical experiments focused on DNA and the often-famous scientists that performed them, along with other more obscure investigations that humbly provided key insights into the nature of DNA and its relationships to the world around it. The present inquiry will consider classic genetic research taking place in the mid-twentieth century, as well as modern research relying on cutting-edge, massively parallel DNA sequencing technologies that now dominate genetic science in the post-genomic era. Following a Popperian approach and attitude, many pointed and complementary efforts will be made to falsify the three specific Buddhist hypotheses under evaluation here. The first phase of the investigation, presented in this chapter, will focus on three general aspects of DNA: the molecule's basic biochemical and biophysical features, its role in "coding" for other biomolecules such as protein, and the nature of DNA's mutation process. The second phase of the investigation, presented in the next chapter, will focus on DNA's role in inheritance and identity.

DNA Structure and Composition

Drama shaped DNA's story, then and now. Half a century prior to the frenzied race between corporate (J. Craig Venter) and public (Francis Collins) powers to provide the first complete draft of a human genome sequence, the 1950s featured a famous trans-Atlantic contest to solve the puzzle of DNA's structure. White male egos were at the center of this competition as well. However, unlike the hypothesis-free duel to define the human genome sequence in the late 1990s, competing scientific hypotheses defined the decades-earlier race to unravel the basic biochemical structure of DNA.

Linus Pauling loved proteins. In the early 1950s, Pauling and collaborators unlocked one of the key molecular mysteries of biology by correctly proposing the alpha helix and beta sheet as the two primary structural features that contribute to the structures and functions of proteins.[3] Pauling's progress resulted from his willingness to challenge conventional wisdom and consider the possibility that a turn of the helix might involve a noninteger number of individual protein subunit molecules (amino acids); for the alpha helix, it was 3.7 subunits per turn. This insight added to the Caltech polymath's growing and diverse list of seminal contributions to science that included the nature of the chemical bond, quantum chemistry, crystal structures of ionic chemical compounds, and the molecular-genetic underpinnings of sickle-cell anemia.

For many years, Pauling's preoccupation with proteins drove him to cling to the hypothesis that these molecules must be where genetic information resides. Although a 1944 paper[4] published by Oswald Avery and colleagues provided strong evidence that DNA, apparently by itself, could transfer new genetic traits between *Pneumococcus* bacteria, up through 1951 Pauling continued to believe that the genes, the heritable units holding the "code" of life, must be present in the proteins:[5]

> I was so pleased with proteins, you know, that I thought that proteins probably are the hereditary material rather than nucleic acids—but that of course nucleic acids played a part. In whatever I wrote about nucleic acids, I mentioned nucleoproteins, and I was thinking more of the protein than of the nucleic acids.

A young postdoctoral scholar working in the Cavendish Laboratory at Cambridge, England, named James Watson acknowledged earlier than most that DNA was the most likely molecular source of genes. Eventually, the broader scientific community—including Linus Pauling—came to agree and the question of "what" gave way to "how" in terms of genetic function. The race was on, and Pauling was behind Watson and others who earlier realized the importance of DNA. Everyone was aware of the basic building blocks of DNA—sugars, phosphates, and nucleotides. It was unknown, however, how all of these pieces fit together to form a three-dimensional structure, and one that could simultaneously serve inheritance and coding functions. In a February 1953 article with Robert Corey,[6] Pauling proposed a triple-helix hypothesis for the structure of DNA, one where the phosphates were neutrally packed into the center of the molecule and the bases were on the exterior. Pauling, as we all now know, was wrong. This scientific blunder dogged him in the years following the famous April 1953 *Nature* paper where the correct structure of DNA was reported by two scientific upstarts working in Cambridge, James Watson and Francis Cricke.[7] After taking liberties with Rosalind Franklin's amazing X-ray crystallography data, Watson and his partner Cricke revealed the details of DNA's now famous double-helix structure with the phosphates and sugars on the outside and bases on the inside. The elucidation of DNA's molecular structure had a domino effect. Subsequent experiments in the following decades—all dependent upon Watson and Cricke's double-stranded model—revealed additional insights into the molecular mechanisms by which DNA functioned in biological heredity and coding for other biomolecules. For Pauling's error, two key factors were identified by a variety of critics that included other scientists, news reporters, Ava Helen Pauling (his spouse), and even Linus Pauling himself: hurry and hubris.[8]

* * *

The double helix became an iconic attribute intimately associated with DNA. Double-stranded mental images promptly arise in the minds of anyone hearing utterances of the word "DNA," at least among those minds benefitting from the requisite education and cultural context. The double-stranded nature of DNA thus provides an ideal first challenge to the three specific Buddhist hypotheses. Is DNA's double-strandedness a permanent, unchanging, intrinsic feature of the molecule? An answer of "yes" would falsify H_s1 and H_s2.

DNA, however, is not always double-stranded. Sometimes it is single-stranded. In fact, anyone handed a tube of double-stranded DNA in water as a solvent could easily make it single-stranded by dropping that tube into a pot of boiling water. Geneticists and biology students working in modern research and teaching labs today routinely heat solutions containing DNA for specific applications requiring the DNA to exist in a single-stranded state. One such application is the powerful polymerase chain reaction (PCR): a laboratory method that creates many millions to billions of copies of a specific target region of DNA. Performing PCR requires certain steps, including one called "denaturation," whereby the scientist intentionally heats the DNA molecules to near boiling points (usually 96°C–98°C). At this temperature, the hydrogen bonds between opposing bases on opposite strands of the DNA break, and the two strands separate from one another to become single-stranded molecules. After making the DNA single-stranded, the next step in PCR, called annealing, involves a lowered temperature (often 50°C–70°C) whereby some regions of the experimental DNA are double-stranded and others remain single-stranded.[9] PCR revolutionized molecular biology and has been applied to multiple biomedical applications ranging from paternity testing to genetic disease analysis to the detection of viruses and other pathogens.

PCR brought its inventor, Kary Mullis, enormous scientific fame and the 1983 Nobel Prize in Chemistry. Mullis earned his PhD in biochemistry from the University of California, Berkeley, in 1973, and later joined Cetus Corporation in Emeryville, California, where he ultimately developed PCR and rose to scientific stardom. Rather than careful implementation of the scientific method or university coursework, Mullis credited contemplation during long California highway drives, on surfboards, and during LSD trips as the keys to his development of PCR. Further, as time went on, Mullis's thoughts and activities moved toward the pseudoscientific; he expressed positive views toward astrology and doubts about climate change, the ozone hole, and HIV as the causal agent of AIDS in his autobiography.[10]

In addition to temperature dependence, the hallmark double-stranded feature of DNA is dependent upon its solvent. DNA is usually studied in water, both inside of cells and when existing inside plastic tubes used by scientists. In 2000, two Massachusetts Institute of Technology chemists published a paper detailing the structure of DNA when taken outside of its native water and moved to various other liquid solvents.[11] They studied short pieces of double-stranded DNA synthesized in the lab (twenty-one base pairs in length) as well as natural DNA collected from calf thymus and salmon sperm. In one unnatural solvent, glycerol, all three types of DNA under study maintained a double-helix structure similar to that observed when the molecule is in water. However, in the three other liquid solvents analyzed (formamide, methanol, and dimethylsulfoxide), DNA's double helix disappeared. Instead, DNA took the shape of unorganized tangles.

The previous examples, however, required human-intentioned, experimental manipulation of DNA molecules. Does DNA's structure ever depart from its iconic double-strandedness when in a "natural" state, free from the effects of human hands? In 1935, a virus that infects bacteria ("phage") was isolated from the sewers of Paris.[12] This phage, named ΦX174, was discovered to have a small, single-stranded DNA genome two decades after its initial discovery.[13] In 1977, ΦX174 became the first organism to have its complete genome (5,386 base pairs in length) sequenced.[14] In 2003, J. Craig Venter added this virus to his list of scientific conquests; Venter led a team that synthesized de novo the entire ΦX174 genome in the lab.[15] Many other single-stranded DNA phages were discovered over the last century, in addition to single-stranded DNA viruses that infect animals, plants, fungi, and other forms of life.[16] Recent advances in environmental DNA sequencing approaches, building off of those pioneered by Venter's 2004 adventures on *Sorcerer II*, suggest that single-stranded DNA viruses are prominent yet vastly understudied components of the world's microscopic ecology. In 2018, for example, a collaborative team of scientists from Japan demonstrated that just one cubic centimeter of ocean sediment contains between one hundred million to three billion single-stranded DNA viruses; in this environment, single-stranded DNA viruses were found to be much more abundant than double-stranded DNA viruses, the former accounting for 96.3–99.8 percent of the benthic total viral assemblages.[17] DNA is double-stranded, but only sometimes and under certain conditions. H_s1 and H_s2 remain supported.

* * *

DNA is made up of molecular building blocks: sugars, phosphates, and nucleotide bases. Biology textbooks teach that the nucleotides provide the "coding" capability of DNA and reside in the safe interior of double-stranded molecules. The books and other basic education materials further explain that there are four nucleotide varieties that make up DNA: adenine ("A"), cytosine ("C"), guanine ("G"), and thymine ("T"). Each of these has a distinctive molecular structure, and hydrogen bonds between nucleotides (A pairing with T, C pairing with G) on opposite DNA strands are essential for the molecule's iconic double-stranded structure. The nucleotides, key determinants of the genetic "code," are widely considered the most important subcomponent features of DNA. In the world of genetics, the nature of a particular DNA molecule under study is often distilled down to the linear order of its nucleotides: the "sequence." Today in the early twenty-first century, many genetic researchers routinely work at the scale of entire genomes and often think about DNA exclusively in terms of the linear order of A's, C's, G's, and T's: the "sequence" of the DNA. Modern genome scientists spend much of their time analyzing enormous amounts of this DNA sequence information, with most of the analytical work done by high-powered computer clusters. Alternating sugars and phosphates make up the exterior "backbone" of double-stranded DNA, as revealed by the model made famous by Watson and Cricke in 1953. The sugar-phosphate backbone is generally characterized as having a protective structural role rather than coding function. Each molecular subcomponent of DNA, in turn, is composed of atomic building blocks such as carbon, hydrogen, oxygen, and phosphorus. Are the A's, C's, G's, and T's of DNA, so prevalent in biology textbooks and scientific minds, static and "self"-defining features of this famous molecule? Are the molecular and atomic building blocks invariant, intrinsic features of DNA molecules? An affirmative answer for either question would falsify H_s1 and H_s2.

DNA molecules residing inside cells experience an ongoing onslaught of chemical perturbation caused by a variety of internal factors. DNA is also subject to molecular change from environmental factors such as cigarette smoke and UV rays from the sun. The production of ATP, the main molecular currency of cellular energy, in the mitochondria of human and other animal cells is accompanied by the production of oxygen free radicals which are capable of chemical interaction with and subsequent modification of DNA nucleotides. For example, oxygen radicals interact with guanines and frequently form 8-oxoguanine molecules; these chemically modified versions of "G" pair with "A" (rather than "C") during DNA replication, and thus can lead to a change in the DNA sequence, also known as a mutation. Oxidation of guanine inside the cell is no rare event: at any given moment, the nuclear DNA inside a typical human cell is estimated to have many thousands of 8-oxoguanine varieties of "G" nucleotides.[18] However, protective DNA repair proteins also circulate inside human nuclei, the intracellular compartments where most of the DNA is housed; these repair factors recognize 8-oxoguanine molecules and convert them back into canonical guanines. Some oxidatively damaged guanines, however, escape this surveillance and lead to mutation events.

Epigenetics is one of the hottest fields of biological research and another avenue of nucleotide change. Biological changes are often considered "epigenetic" when there are functionally relevant changes to DNA, or proteins that interact with DNA, that do not change the primary DNA sequence. Epigenetic changes are increasingly implicated in diverse processes of life and the human experience, ranging from embryonic development to aging to stress and depression.[19,20] Epigenetic changes can thus play roles in *duhkha*. Cytosine methylation (addition of a CH_3 group) by enzymatic proteins called DNA methyltransferases provides one well-understood example of epigenetics. In humans, three different DNA methyltransferase genes are present in the genome and encode enzymes that transfer methyl groups onto cytosine nucleotides, often at "CpG" sites (sequences where a "G" follows the "C").[21] The methylation status of DNA impacts its interactions with other biomolecules. In general, regions of DNA with low levels of methylation undergo transcription (the molecular process that produces RNA, which in turn leads to the production of protein) at relatively high rates, whereas highly methylated DNA is transcribed at relatively low rates. Cytosine methylation is a critical and essential functional aspect of mammalian life; like humans, mice encode three DNA methyltransferase genes, and experiments focused on mutant and dysfunctional versions of these genes reveal that all of them are required for a mouse to complete embryonic development.[22,23]

Methylation, however, also has mutational consequences. Methylated cytosines undergo deamination (loss of a NH_2 group) at higher rates than typical "C" nucleotides, a process that results in the formation of uracil ("U"), a nucleotide usually associated with RNA instead of DNA. "U" bases pair with "T" bases, whereas "C" (in methylated and unmethylated forms) pairs with "G." Thus, like guanine oxidation resulting from DNA damage, methylated cytosines produced by DNA methyltransferase enzymes can lead to mutation events. Although cytosine methylation is the most widely studied and best understood form of epigenetic nucleotide modification, recent research reveals that cytosines can also be modified into other ways (such as through the addition of hydroxymethyl groups) and that adenine is also subject to methylation in some species.[24]

DNA sequences—strings of A's, C's, G's, and T's—that appear in genetics textbooks and the computer screens of genome biologists offer convenient though misleading shorthand descriptors for the nature of nucleotide existence. The nucleotides of DNA inside cells are in constant existential flux, resulting from diverse forms of chemical damage (often followed by enzyme-induced repair) and epigenetic modifications. These changes can often lead to mutations. The four letters that make up a DNA sequence string, however, imply a sense of stability—perhaps a sense of "self" in the minds of some scientists—in nucleotide chemical structure for what is in truth a highly unstable and impermanent situation. H_s1 and H_s2 thus continue to remain supported.

Although the dynamic nature of nucleotides supports H_s1 and H_s2, the molecular structure of DNA offers another avenue to falsifying these hypotheses: the sturdy backbone. The sugar-phosphate backbone, composed of alternating deoxyribose sugar and phosphate molecules on the exteriors of double-stranded molecules, is often portrayed as a stable agent of protection that shields the more chemically labile nucleotides on the inside of double-stranded DNA. But, in reality, backbones break. If a scientist heats DNA in water solvent past the boiling point (above 100°C), beyond the near-boiling temperatures used to make DNA single-stranded in PCR experiments, the bonds between sugars and phosphates in the DNA backbones will break apart. UV radiation also causes single- and double-stranded DNA breaks. Further, at least in some bacteria, the backbone is also subject to natural epigenetic change, such as the enzyme-induced phosphorothioate modification whereby an oxygen atom in the sugar-phosphate backbone of DNA is replaced by a sulfur atom.[25] This modification makes the bacterial DNA more resistant to phage DNA integration. Thus, the backbone of DNA is also subject to change, and it can consist of varying molecular components. H_S1 and H_S2 remain supported.

* * *

Since the molecular building blocks—the nucleotides, sugars, and phosphates—of DNA have thus far failed to falsify H_S1 and H_S2, it is now time to dig deeper to the level of atoms. Molecules are made of atomic building blocks. For example, each sugar in DNA's backbone is made up of interconnected carbon, oxygen, and hydrogen atoms. How stable is the atomic composition of DNA?

Hydrogen and other atoms occur in different forms, called isotopes. Most hydrogen atoms are composed of two main subatomic parts: a single proton comprising the positively charged atomic nucleus, and a single negatively charged electron that orbits the nucleus. In some cases, the hydrogen nucleus is composed of a proton particle joined by an uncharged neutron particle. Such "heavy" hydrogen atoms are referred to as "deuterium." "Protium" is used to refer to "light" hydrogen atoms that lack neutrons. Water molecules composed of oxygen atoms bound to two deuterium isotopes are referred to as "heavy water"; water molecules containing typical protium are called "light water."

Atomic exchange experiments take advantage of the potential for molecules to exist in varying isotopic states. In this simple but powerful approach, focal molecules (DNA or proteins, for example) are initially kept in light water and then transferred into heavy water for subsequent molecular investigation. After an incubation period, time when the molecules under study are immersed in the new heavy-water solvent, the molecules are then collected and analyzed. When atomic exchange occurs, the original hydrogen isotopes (protium in this example) that were originally part of the molecule have the opportunity to undergo "exchange" with the new hydrogens present in solvent water molecules (deuterium in this example). The procedure can also be done vice-versa, starting with deuterium and moving into a protium solvent. No enzyme or any other external force is required to mediate the atomic exchange; it occurs through the natural interactions between solutes (the molecules under study) and the solvents (molecules in which they dissolve, with which they interact).

Density gradient tubes were used in the 1960s to detect evidence of atomic exchange; nuclear magnetic resonance and mass spectroscopy are the preferred tools today. These scientific instruments reveal the occurrence of atomic exchange between target molecules and solvents in such experiments. Protein molecules have been the most prevalent and favorite study subjects of atomic exchange experiments; in many cases, exterior regions of folded protein molecules are accessible to the solvent and undergo rapid atomic exchange, whereas internal regions are inaccessible to the solvent and rarely undergo exchange. Atomic exchange studies were important to illuminating common structural features of proteins that circulate in cellular cytoplasm, such as their hydrophilic exteriors that interact with water and other molecules and the hydrophobic cores that play important roles in determining protein structure.

A 1962 hydrogen exchange study targeted DNA rather than proteins as the focal molecule.[26] DNA molecules initially composed of deuterium were moved to a protium solvent and then analyzed. The results revealed that complete hydrogen exchange—all of the DNA's original deuteriums replaced by the solvent's protiums—occurred in less than three minutes. The timing could not be made more precise, because three minutes represented the time it took to set up the experiment in the first place! In the snap of a finger, light atoms from the solvent completely replaced all of the DNA molecule's original heavy hydrogens. Likewise, the heavy hydrogens originally composing the DNA molecule became part of the water solvent. The DNA molecule's atomic identity and the identity of the water blurred, changing faster than the scientists could look. This atom's-eye view of DNA is consistent with H_s1 and H_s2. The results of this exchange experiment, performed more than fifty years ago, offer evidence for a mutually interdependent coexistence between the solute under study (DNA) and its solvent (water). Thus, this experimental outcome is also consistent with H_S3.

Genetic Codes

Scientists cracked the genetic code in the 1960s. Earlier research had demonstrated details on the structure of DNA, which provided clues to the mechanism by which it coded for protein molecules, and evidence that this molecule was the home for the heritable "genes," the individual genetic units encoding biological traits. However, the precise molecular interactions required for the "information" in DNA to become expressed as proteins and other biological traits remained mysterious. Many complimentary laboratory investigations were required to piece it all together, though one essential experiment provided noteworthy links to the present analysis.

In the early 1960s, a trio of molecular biologists convened at Caltech in Pasadena, California, to collaboratively seek new insights into the genetic code. Matthew Meselson, a key contributor to our understanding of how DNA replicates, was stationed at Caltech and served as the host. Two European scientists joined Meselson in Pasadena for the project: Sydney Brenner from Cambridge (who eventually became a Zen aficionado) and Francois Jacob from the Pasteur Institute (who eventually granted a PhD in molecular genetics to Matthieu Ricard, now a Tibetan Buddhist monk). At the time of this trio's collaboration, evidence was mounting that there must be some kind of intermediary molecular "go-between" linking DNA to protein production in the ribosome (the place of protein synthesis inside cells), but the identity and nature of that "messenger" molecule remained unknown.

Brenner, Jacob, and Meselson conducted an experiment with *E. coli* and an infecting phage as tools to investigate the mystery messenger. They used a "heavy versus light water" hydrogen exchange approach, as previously described. The trio took advantage of the fact that *E. coli* cells, when infected by phage, halt DNA synthesis for approximately seven minutes, whereas ribosomal protein production continues uninterrupted. During this seven-minute lag associated with phage infection, protein production switches from "host" bacterial proteins to those encoded by the phage DNA. *E. coli* cells were initially grown in a "heavy" growth medium, and then infectious phage particles were added. After infection, the bacteria were moved to a "light" medium, resulting in all subsequently synthesized constituents being of the light variety. Analysis of the newly synthesized biomolecules revealed the presence of a transient RNA molecule as the prime candidate for the key molecular intermediary. This newly synthesized RNA was found to have a nucleotide composition highly similar to that of the phage DNA, suggesting that the RNA was a simple molecular copy (though single-stranded rather than double-stranded) of the DNA.[27] This experiment shed key light on the final missing "middle piece" in biology's mission to connect the molecular dots between DNA and protein. This type of biomolecule eventually became known as "messenger RNA," a single-stranded molecule nearly the same as DNA in every way, except for the presence of uracil ("U") instead of thymine ("T") and a slightly different chemical nature to the sugar in the backbone (ribose instead of deoxyribose). This and other research led to the formulation of the "Central Dogma" of molecular biology: DNA undergoes *transcription*, which produces a single-stranded messenger RNA molecule that is complementary to the coding DNA strand; the messenger RNA in turn undergoes *translation* into a linear chain of amino acids, which eventually fold up to form a functional protein.

Many additional experiments in the early 1960s unraveled additional details of Central Dogma molecular mechanisms, such as the famous experiment conducted by Francis Crick, Sydney Brenner, Leslie Barnett, and R. J. Watts-Tobin which revealed the "triplet" nature of codons: a single codon was defined as three consecutive nucleotides along a strand of DNA that encoded a single amino acid in a given protein.[28] Linear strings of consecutive codons in a gene encoded correspondingly linear series of amino acids in a protein. A triplet genetic code, combined with four possible nucleotide states at each codon position, implicated sixty-four total possible codon sequences. The code was also characterized as degenerate: some amino acids were found to be encoded by many codons. This first-discovered genetic code for *E. coli* was found to encode twenty different amino acids from sixty-one different codons, along with three codons serving as stop signals which terminate translation. Research in the years that followed revealed that the genetic code of *E. coli*, the specific assignments of amino acids to certain triplet codon sequences in this bacterial species, was shared in additional fungi, animals, plants, and in humans.[29] This observation led to the characterization of the code as "universal," a moniker that continues to cling to descriptions of the genetic code today. Is the genetic code truly invariant in all forms of life? An answer of "yes" would falsify H_s1.

Although the genetic code continues to be widely described as "universal," it has been known for decades that the code varies. In fact, the code is not even constant within a human body or even a human cell! In 1979, a team of three Cambridge scientists analyzed DNA from the mitochondria of humans and cows; they discovered that two codons had evolved different coding functions in these organelles as compared to the "universal" code used in human and cow nuclei.[30] Their experiments showed that, in mitochondria, UGA codons encoded tryptophan amino acids instead of serving as a stop signal, and AUA codons encoded methionine instead of isoleucine amino acids. Shortly after this finding in mammalian mitochondria, genetic code variations were soon discovered in the mitochondria of additional species of fungi and invertebrate animals.[31] Further research revealed additional code modifications in viruses and other microorganisms, including genetic code expansion to include additional amino acids such as selenocysteine and pyrrolysine.[32] Recently, scientists have taken to tinkering with and "reprogramming" the genetic code in the lab to create new varieties of biomolecules.[33] Thus, DNA's so-called universal genetic code is not a truly invariant and fixed feature of life systems, a reality that is consistent with H_s1.

Simplified descriptions of how the genetic code works, the Central Dogma, usually follow the reductionist causal chain approach: DNA is transcribed into messenger RNA (transcription); messenger RNA is translated into protein (translation). However, such abridged descriptions gloss over the numerous interactions required for the production of a given protein molecule. The key molecular events of transcription and translation are prominent in the undergraduate curriculum of future nurses, doctors, pharmacists, lab technicians, agricultural scientists, and academic biologists. The steps of these central "information processing systems," however, are usually described from the "bottom-up" and unidirectional view of cause and effect (DNA → RNA → protein), a description challenged by Denis Noble in his systems approach to understanding how life works. In works such as *The Music of Life*, as previously discussed in Chapter 5, Noble presented numerous examples that counter the reductionist causal chain characterization of the Central Dogma. Many of his examples focused on the effects of higher-level biological processes on those at lower levels, such as the effects of hormone production on gene expression. Further, Noble highlighted the fact that proteins are required for transcription as another example in support of a critical view of the unidirectional, reductionist characterization of the Central Dogma. Noble recognized a convergence between his systems view of life and *anatman*; his decades of work in this area offer numerous avenues of support for H_s2.

Does the genetic code offer any evidence of linear cause and effect that is inconsistent with the "mutual" aspect of *pratityasamutpada*? An initial inspection of the process of transcription, the production of a messenger RNA molecule from a template DNA, suggests a unidirectional process in potential conflict with *pratityasamutpada*, and thus also H_s3. In transcription, DNA appears to be a one-way "cause" of the "effect" of RNA production, at least on the surface. Does the process of transcription have any effects on the DNA molecule? In the initial phase of transcription, a collection of proteins (the RNA polymerase and a variety of other proteins called transcription factors) bind to specific regions of DNA called promoters and then unwinds a fourteen–base pair segment of the DNA called the "transcription bubble." Then, the production of an RNA molecule by the RNA polymerase and associated molecules begins through base-pairing interactions between one of the single-stranded DNA regions (the template strand) and free ribonucleotides. The elongation phase then involves the transcriptional machinery rapidly moving along and unwinding the DNA molecule, with regions behind rewinding along the way and continuously adding successive ribonucleotides onto the growing RNA chain through additional subsequent base pairing interactions with the template DNA strand. Transcription terminates once the complete single-stranded RNA molecule is completed, and the RNA polymerase and other transcriptional machinery dissociate from the DNA, which returns to a completely double-stranded state. Thus, the process of transcription has effects on the DNA molecule; it is not merely a passive bystander. RNA polymerases and other proteins directly bind the DNA, segments transition from double-stranded to single-stranded states, and the nucleotides of the template strand undergo interactions with ribonucleotides to create the growing single-stranded RNA chain. The DNA, RNA, and protein are all mutual causes and effects of one another's changing states throughout the molecular orchestra commonly called transcription, a situation highly consistent with H_s3.[34]

Random Mutation

Chemical changes in DNA molecules, such as oxidation of guanines or methylation of cytosines, can lead to changes in the DNA sequence, mutation events. Mutations result naturally from a variety of forces, including UV radiation, cigarette smoke, and errors during DNA replication. Humans can intentionally create desired mutations in the lab, with targets ranging from the genomes of E. coli inhabiting Petri plates to the DNA inside human embryos. The mutation process is a fundamental topic of study in biology; mutations are causal agents contributing to all cancers and genetic diseases, and they are the fuel for evolution in all forms of life. From an evolutionary perspective, mutations are of central importance because they are the source of all genetic variation in natural populations. Mutations are heritable varieties of DNA change; any evolutionary advantage or disadvantage offered by a given mutation will have the potential to continue in future generations that continue to bear that DNA sequence variant.[35] Given the centrality of mutation in the evolutionary process of life, the nature of the mutation process has preoccupied the minds of many geneticists and evolutionists over the last century.

Alongside the frenzy of research revealing insights into the structure of DNA and nature of the genetic code during the mid-1900s, additional scientific attention centered on the question of whether mutation was a random process or one that was "directed" by the environment. For example, does the presence of a particular antibiotic in the growth media of E. coli cells, initially sensitive to that antibiotic, somehow direct mutations to the specific genes associated with antibiotic resistance? Or does the mutation process proceed in a fashion that is random in this sense, with mutations events at one site being equally likely as mutations at any other site, blind to the potential impacts that might result as a consequence?

In the early 1950s, Joshua and Esther Lederberg—a husband-and-wife scientific duo working at the University of Wisconsin—set out to investigate the nature of mutation. Like many other scientists investigating DNA at the time, they relied on an E. coli and infecting phage laboratory system. This pair developed a replica plating experimental approach to distinguish between the competing hypotheses of random versus directed DNA mutation. These experiments began with an agar Petri dish already populated by many dozens of visible and distinct E. coli colonies randomly distributed across the plate. Each tiny, circular colony of cells originated from a single progenitor cell which, over the span of approximately twelve to twenty-four hours in a laboratory incubator, multiplied to billions of cells to form the naked-eye-visible colony. This starting plate was called a "master plate" in replica plating experiments. In the Lederbergs' experiments, the master plates contained bacteria growing on basic growth nutrients, but no phage. In the next step, the scientist gently pressed a sterile velvet pad—stably secured to a paddle with a handle occupying the scientist's hand—onto the bacterial colonies-containing agar of the master plate. Many representative cells, thousands to millions, from each colony transferred over to the velvet pad, in the same relative orientations they earlier occupied on the master plate. The velvet was next pressed down onto a new plate, called the "replica plate." Representative cells from each of the original colonies on the master plate became new residents of the replica plate. After some time in the incubator, the cells transferred to the replica plate grew up to form new visible colonies. The specific spatial ordering of colonies, originally present on the master plate, was preserved on the newly created replica plates.

The Lederbergs reasoned that, if mutations were random, then prior exposure to some stressor, such as a phage, would not be a prerequisite for mutations causing phage resistance in *E. coli*. Under this hypothesis, some of the bacteria making up colonies on the master plate (where phage is absent) might already have DNA that randomly experienced mutation in genes that confer phage resistance. Further, those colonies that happened to contain bacteria with such mutations would be expected to often contain many such bacteria, due to the many successive rounds of cell division required to give rise to the visible colony. The Lederbergs predicted that a given master plate colony that was also able to produce growing colonies on a replica plate that contained phage (that would kill bacteria lacking phage resistance mutations and prevent colony formation) would be able to do so repeatedly, on many different phage-containing plates. They reasoned that if the mutations responsible for phage resistance were already there in the master plate colony cells, then some representative resistant bacteria should make their way from the master plate to the velvet to the phage-containing replica plate and then grow to form a visible colony. Because the mutation was there to begin with in the master plate colony, the pattern of colony growth (i.e., the particular master plate colonies that do or do not grow on the phage-containing replica plate) should be repeatable from replica plate to replica plate. Such an experimental outcome would be consistent with a hypothesis of random mutation and inconsistent with a hypothesis of directed mutation.

What about the alternative hypothesis? Somewhat ironically, a random and unrepeatable pattern of colony growth on replica plates would oppose the hypothesis of random mutation. This kind of results would indicate that phage resistance mutations did not preexist in the master-plate bacterial colonies and thus might have been induced only after exposure to the phage on the replica plate. Observations of random and varied occurrence of phage-resistant colonies, differing from replica plate to replica plate, would suggest that mutations conferring phage resistance might require exposure to the phage. This sort of hypothetical mutation process would be described as *induced* or *directed* by the phage environment present on the replica plate in this experiment and absent on the original master plate. Such an experimental outcome would be inconsistent with a hypothesis of random mutation; instead. it would support a hypothesis of directed mutation.

In their Wisconsin lab, the Lederbergs repeatedly observed that, for a single given master plate, the pattern of colony-specific phage resistance repeated from replica plate to replica plate.[36] This result was consistent across many different sets of master plates. This experimental result, coupled with additional complimentary research findings such as the Luria-Delbruck fluctuation experiment,[37] led to a general consensus in the genetic community that mutation was a random process, not one "directed" by the surrounding environment.

The random nature of the mutation process, initially conceived in a rather narrow sense relating to the concept of directed mutation previously discussed, conceptually expanded in the minds of many molecular and evolutionary biological thinkers such that mutation became characterized as a *purely* random process. For many biologists, mutation became conceptually and mathematically treated as totally random, with any particular nucleotide equally likely to undergo mutation as any other. In the development of modern evolutionary genetic theory over the last century, these completely random mutations became a foundational substrate upon which the nonrandom process of natural selection acted. Is the randomness of mutation an inherent feature of DNA? Evidence in support of this would provide paths to falsifying H_S2 and H_S3.

* * *

The Dalai Lama's 2002 Mind and Life Dialogue, previously discussed in Chapter 5, focused on the nature of life, and the role of mutation in the evolutionary process became a point of disagreement between the monk and the scientists. The biologists at the event described the mutation process as one which was "totally random," following the general scientific thinking prevailing in evolutionary genetic circles. This caricature, however, conflicted with the Dalai Lama's Buddhist outlook on nature. In *The Universe in a*

Single Atom, the monk questioned whether random mutation was "best understood as an objective feature of reality or better understood as indicating some kind of hidden causality."

The Dalai Lama had no problem with the idea that DNA changes (consistent with H_s1), but doubted the idea that this change occurred in a random, willy-nilly fashion, free from the broader context of cause and effect expressed in Buddhist thought (i.e., in accord with *pratityasamutpada*). The idea that mutation is purely random, with any particular nucleotide being just as likely to mutate as any other regardless of context or interactions, opposed the monk's Buddhist view of reality and the universe.

The molecular mutation process, in fact, is nonrandom in many ways. First, there exist well-understood mutation hotspots that occur in genomes. In humans and other mammals, "CpG islands," regions of DNA with high densities of CpG dinucleotides and thus high levels of methylated cytosines, are regions that experience high mutation rates relative to other parts of the genome.[38] As previously discussed, methylated cytosines undergo deamination to uracil at high rates and thus lead to C:G to T:A base substitution mutation events. Repetitive DNA sequences offer a second example of mutation hotspots. Homopolymeric nucleotide runs, stretches of DNA composed of repeating single base pairs, undergo insertion-deletion mutations (usually the insertion or deletion of one of the repeating base pairs) at rates approximately one hundred to one thousand times higher than nonrepetitive regions of DNA.[39] This increased rate results from a higher-than-usual incidence of DNA polymerase slippage events during replication at these repetitive sequences. Mutations in similar repeat regions of DNA underpin certain human genetic diseases such as Huntington's disease and Fragile X syndrome. A third example derives from a 2012 analysis of mutation in *E. coli* DNA, which revealed a high degree of heterogeneity in mutation rates across genome space.[40] In fact, coding regions of DNA (those acted on by RNA polymerases to produce messenger RNAs) were found to experience lower mutation rates as compared to noncoding DNA in the *E. coli* genome. Another study published in 2012 revealed extensive mutation rate heterogeneity in human cancer cells.[41]

The Dalai Lama was thus wise to doubt the scientists' overly simplified characterization of mutation as a purely random process. It is not. Any nucleotide in the genome is susceptible to mutation, though they are not all equally susceptible. Mutations, when they happen, result from contextual interacting factors such as UV radiation, oxidative modification of nucleotides from oxygen free radicals in the cell, and errors during the DNA replication process. The mutation process is dependent upon the conditions in which the DNA molecule finds itself. It remains true that there is no evidence for any kind of "directed" mutation that would promote DNA changes in the genes that would be favorable; in this very specific and relative sense only, evidence points to mutation as a random process in this regard. Human intention, however, is capable of directing mutation to desired locations, such as the controversial editing of DNA in twin human embryos by the Chinese scientist He Jiankui in 2018. Mutation, thus, is not a purely random process and depends deeply on context and interactions, consistent with H_S2 and H_S3.

Review

Three different facets of DNA—its biophysical structure and composition, coding functions, and mutation process—have now been investigated. Pointed efforts to falsify the three specific Buddhist hypotheses based on well-known features of DNA expected to pose challenges have fallen short. H_S1 predicted that all features of DNA would be impermanent. However, its iconic double-strandedness was found to disappear with increasing temperature, and its biochemical composition was found to be variable in a variety of circumstances. Further, the so-called universal genetic code of DNA was found to be not so universal. H_S2 predicted that DNA lacks inherent features characteristic of a "self" nature for the molecule. Its double-stranded nature was found not only to be impermanent but also to depend upon the form of life in which the molecule is found, with single-stranded molecules actually prevailing among the countless phage particles occupying ocean sediments. Hydrogen exchange experiments revealed an essentially instantaneous atomic exchange between DNA and its water solvent, blurring the lines of identity between solute and solvent.[42] The hallmark and apparently inherent "random" nature of DNA mutation was discovered to be an exaggerated oversimplification. H_S3 predicted that DNA exists and functions in a mutual cause-and-effect framework as expressed through the Buddhist doctrine of *pratityasamutpada*. A careful inspection of the molecular events underpinning transcription revealed a cause-and-effect scenario whereby DNA was no mere bystander, but experienced effects from the molecular transcription process (e.g., unwinding into single-stranded regions, interacting with RNA polymerase and other proteins) as well as contributing cause to the production of RNA molecules. No true "first cause" or "single cause" associated with DNA has been uncovered in the tour through decades of research focused on this molecule. For example, the synthesis of a new DNA molecule through the replication process in cells requires myriad pre-existing biomolecules (e.g., free nucleotides, DNA polymerases, etc.) that come together as "multiple causes" to orchestrate the synthesis of new DNA molecules. The first phase of the investigation has concluded, and the three specific Buddhist hypotheses remain supported thus far.

8

Identities

Mendel was a monk. Born in Heinzendorf bei Odrau in 1822, now part of the Czech Republic (then part of the Austrian Empire), the boy given the name Johann Mendel at birth performed the family farm duties of a peasant youth until the age of eleven. At that time, a local headmaster saw in Johann an exceptional intelligence and enthusiasm for learning; the headmaster recommended young Johann for admittance to the gymnasium in Tropau. Mendel's family hesitantly obliged, despite his father's disability (he was injured in a logging accident) and the financially strained status of his family. Mendel graduated six years later with honors, but along the way he had his first incident of "mental incapacitation"; today this would be called clinical depression. These depressive episodes would recur in Mendel's future life. After his graduation, Mendel continued his education at the Philosophical Institute of the University of Olmutz, where he studied philosophy and physics. Here he excelled, especially in math and physics. A severe bought of depression, however, required him to take a hiatus from his studies and delayed his graduation from the university. In Olmutz, Mendel met Johann Nestler, the dean of natural history. Nestler, a notable biologist, was preoccupied by the rules of heredity. He had an established research program focused on animal and plant breeding; Nestler was an important early influence on Johann Mendel, then a young student who would later come to be known as the Father of Genetics.[1]

At the age of twenty-two, Johann decided to become a monk, much to the chagrin of his father who would have preferred his son to return home and take over the family farm. Mendel joined the Augustinian order at the St. Thomas Monastery in Brno and began theological studies. This life transition also brought Johann a new name: Gregor.[2] At the monastery, Mendel gained access to a huge library, as well as the teachings and thoughts of eminent thinkers that surrounded him. The monastic community included many well-known philosophers, mathematicians, and botanists. Gregor rose quickly through the ranks and gained his own parish at the age of twenty-six.

The stress of these new clerical duties, however, would prove to be too much for Mendel; his mental illness reached new levels and rendered him bed-bound. The head of St. Thomas, abbot Cyril Napp, thought that Mendel required a pivot in his path; in an 1849 letter to bishop Schaffgotsch, Napp remarked:[3]

> He (Mendel) is very diligent in the study of sciences but much less fitted for work as a parish priest, the reason being he is seized by an unconquerable timidity when he has to visit the ill.

Mendel next bounced around substitute schoolteacher postings and also pursued advanced scientific studies at the Royal Imperial University in Vienna. Twice Mendel attempted the teaching state certification examination to become a regular full-time teacher; twice he failed. Though he was known to be a vivid and engaging presence in the classroom as a substitute teacher and to possess a great intellect, severe test anxiety prohibited the monk from gaining the certification necessary to become a teacher. A bright silver lining accompanied these failed certification examinations: they led Mendel to focus more keenly on pursuing scientific knowledge at the St. Thomas Abbey. For eight patient years, starting in 1856, Gregor Mendel diligently worked in his pea garden.

The monk quietly performed many experiments on thousands of pea plants, with the initial humble intention of developing new pea color varieties and to examine the effects of plant hybridization.[4] These experiments would end up unraveling the basic laws of inheritance. Mendel's selection of peas turned out to offer key advantages to his genetic analysis, some known and others unknown to him at the time. The monk deliberately chose the pea plant system due to the species' ability to self-fertilize and the ease of growing many plants in small spaces. He also applied a rigorous quantitative approach, analyzing patterns of trait inheritance observed among hundreds of plant progeny. Unbeknownst to Mendel, however, most of the genes associated with the pea traits he studied were on separate chromosomes; this made the analysis and interpretation of crosses very simple and straight-forward.[5] Mendel's pea plant research wrapped up in 1863; he completed his analyses and prepared the scientific manuscript over the next two years. Mendel submitted his seminal work that revealed the laws of heredity as a paper, "Versuche uber Pflanzen-Hybriden" (Experiments in Plant

Hybridization), to the Proceedings of the Natural History Society of Brno in 1865, where it was ultimately published in 1866. Mendel's landmark contribution to science, however, would go essentially unnoticed and unappreciated for decades.

In the "rediscovery" of Mendel's results during the early 1900s, many years after the monk's death, three European botanists (Erich Tschermak in Austria, Hugo de Vries in the Netherlands, and Carl Correns in Germany) each independently replicated the monk's results. The three plant biologists, however, discovered that the theory of heredity had already been published in 1866 by Mendel; each botanist realized this only after his own research toward this end had already been completed. Each scientist went to great lengths to show that he had read Mendel only after conducting his own experiments and reaching his own conclusions.[6] It is speculated that De Vries did not originally intend to mention Mendel in his paper, doing so only after learning that the other two botanists had acknowledged Mendel's work.[7]

In addition to the drama surrounding the issue of scientific primacy, the ego boost that comes with doing something "first," Mendel's data also attracted criticism from Ronald Fisher, a central figure in statistics and its application to biology. The now famous "goodness of fit" paper by Fisher suggested, from a statistical point of view, the possibility that some of Mendel's data were falsified.[8] Fisher showed that the monk's observed results were too close to the theoretical expected values; there were too few random errors, and his math was too precise. Fisher suggested a 1 in 2,000 chance that Mendel's experimental results reflected reality. Though we can never know for sure, most (including Fisher) doubted that Mendel intentionally falsified his data to make them align with theoretical expectations and hypotheses. Instead, factors such as the throwing out of data from trials not aligning with other results and unconscious bias have been proposed as possible explanations for Mendel's "too good to be true" pea plant data.

Mendel dwelled in many forms of *duhkha* throughout his life, from failures to meet his father's expectations to test anxiety to the administrative headaches later in life when he became a monastery head abbot. The monk also often felt underappreciated among scientific circles, frustrated that his contemporaries were unable to recognize the impact and relevance of his pea plant results. He did find solace and peace, however, working in his pea garden. Mendel famously stated:[9]

I have experienced many a bitter hour in my life. Nevertheless, I admit gratefully that the beautiful, good hours far outnumbered the others. My scientific work brought me such satisfaction, and I am convinced the entire World will recognize the results of these studies.

* * *

Mendel's seminal findings, after their rediscovery, established the baseline framework for the field of genetics, the study of how traits are transmitted from one generation to the next. Decades later, findings by other scientists such as Oswald Avery and the iconic 1952 "blender" experiments by Martha Chase and Alfred Hershey revealed DNA as the carrier molecule of genetic information,[10] from one generation to the next. Ever since, DNA has gained a reputation as being *the* molecule of heredity and a central determinant of an individual's identity. DNA's role in inheritance and identity has many practical benefits in human society, such as paternity testing and forensic applications. DNA has been used as evidence to link suspects to crimes in legal proceedings, and to exonerate those wrongly accused through the works of nonprofit justice agencies such as the Innocence Project. DNA's reach extended into other areas of biology, such as in phylogenetic applications where patterns of DNA sequence similarity are used to infer evolutionary relationships among different organisms or groups of organisms. This approach gave rise to an entirely new way to delineate boundaries between different species, the "phylogenetic species concept."[11] DNA became a "barcoding" tool to identify various forms of life,[12] and high-throughput sequencing technologies have enabled environmental DNA-based monitoring of ecosystems, "eDNA" applications.[13]

The perception of DNA as a key identity definer in broader society has been reinforced by the advent of affordable commercialized genetic testing applications offered by companies such as Ancestry.com, 23&Me, and African Ancestry. "It's in my DNA" and similar phrases have arisen to describe those aspects of one's being that are inherent to one's "self" nature. For example, Compton rapper and 2018 Pulitzer Prize winner Kendrick Lamar relied on this identity-defining aspect of the molecule in his song "DNA":[14]

> I got loyalty, got royalty inside my DNA
> Cocaine quarter piece, got war and peace inside my DNA
> I got power, poison, pain and joy inside my DNA
> I got hustle, though, ambition, flow inside my DNA

The second phase of evaluating the three specific Buddhist hypotheses will focus on the role of DNA in inheritance and identity (i.e., "self") determination. Is DNA the exclusive unit of heredity, the sole carrier of genetic information? Are all the copies of my DNA, in the trillions of cells in my body, the same? What distinguishes "my" DNA from others' DNA?

Heredity

DNA is known as the molecule of inheritance. In humans, adult individuals are composed of trillions of cells, the vast majority of which are "diploid": each diploid cell contains twenty-three pairs of DNA-containing chromosomes,[15] with one member of each pair typically coming from either the mother or father. The gonads (ovaries, testes) of adults, however, also produce haploid gamete cells (eggs, sperm), each containing a single representative chromosome for each of the twenty-three pairs in diploid cells. When sperm and eggs fuse during fertilization, a diploid cell called a zygote results, which constitutes the earliest phase of what is conventionally considered a new organism, the next generation. The single-celled zygote undergoes numerous successive rounds of cell division, the process called mitosis, thus forming a multicellular entity known as an embryo up until the ninth week after conception, after which the still developing entity goes by the name "fetus." After birth, the fetus becomes known as a baby. The baby develops into an adult, and then eventually becomes capable of producing haploid gamete cells of its own that might contribute to the production of new zygotes, the next generation. This conceptual template for reproduction in humans and other animals, underpinning modern genetic thinking and education, was constructed from a generalized fusion of Gregor Mendel's experiments and the chromosomal theory of heredity formulated by the American biologist Walter Sutton and the German biologist Theodor Boveri in 1902.[16]

Despite the paradigm of inheritance outlined earlier—where sexual diploid organisms produce haploid gametes that fuse to form new diploid organisms in the next generation—there are other well-known mechanisms of reproduction that deviate from this template. The microbial world offers countless counterexamples, ranging from the clonal propagation of haploid bacteria through cellular fission to the replication of single-stranded DNA virus molecules inside their bacterial hosts. In the world of animals, there are also many examples of species that reproduce through alternative means. For example, genetic evidence for parthenogenesis—the production of offspring by females without male fertilization—was reported for Komodo dragons in 2006[17] and hammerhead sharks in 2007.[18] Transitions to asexual reproduction are sometimes associated with changes in the numbers of chromosome sets (ploidy); one example is the occurrence of triploidy (three sets of chromosomes) in *Heteronotia binoei*, an Australian lizard species with parthenogenetic females.[19] Despite the dynamic and evolutionarily labile chromosome content shifts between haploid and diploid and higher ploidy states in different forms of life, the DNA molecule remains at the center of the heredity process. But is DNA always the molecule of heredity? An answer of yes would challenge $H_S 1$.

* * *

Many viruses utilize DNA as their genetic material, but others use RNA. The infamous human immunodeficiency viruses (HIV) offer one example; each HIV particle contains a single-stranded RNA molecule genome. Upon infection of a host human cell, the RNA molecule undergoes a process of reverse transcription that ultimately produces a double-stranded DNA molecule that becomes integrated into the nuclear genome of the host human cell. Viruses that replicate in this fashion are referred to as retroviruses, and they rely on a viral RNA-encoded reverse transcriptase enzyme to produce new copies of their genomes this way. Diverse retroviruses pose threats to many animal and plant species; in addition to HIV, additional retrovirus examples include Hepatitis B and Cauliflower mosaic virus. In the host genome-assimilated state, the HIV-derived DNA can lie dormant for extended periods of time, or produce new infectious HIV particles depending upon the conditions inside the host cell. Without anti-retroviral treatments, the RNA genome of HIV can transmit from infected mothers to offspring at a frequency of 15–30 percent during gestation or labor; virion transmission can also occur during breastfeeding at 15–20 percent rates.[20]

Other kinds of RNA viruses replicate in a fashion without any sort of DNA intermediary. Hepatitis C viruses infect ~2 percent of the global human population, and harbor RNA genomes that replicate inside human cells directly into new RNA molecules, through the activity of a virus genome-encoded RNA-dependent RNA polymerase enzyme. Individuals infected by Hepatitis C sometimes remain unaware of the virus's presence for many years; such chronic infections usually eventually progress to cause cirrhosis or liver cancer. Perinatal transmission of Hepatitis C from infected mothers to babies occurs in humans, though at <10 percent frequencies.[21] Other examples of RNA viruses that replicate without DNA intermediaries include the coronavirus associated with the 2002–2003 outbreak of severe acute respiratory syndrome (SARS), along with the 2019 novel coronavirus (SARS-CoV-2) responsible for the COVID-19 pandemic that, at the time of this writing, has killed more than half a million people on the planet.[22] Thus, whether considered as the genetic material of viruses themselves, or as heritable units capable of transmitting across generations in infected host organisms, the genomes of RNA viruses provide one example of DNA-free inheritance.

The world of epigenetics provides additional examples of DNA-free heredity, in cellular life forms. In addition to the incidence of chemical DNA modifications (e.g., cytosine methylation) discussed in the previous chapter, various mechanisms of RNA- and protein-based inheritance are known to occur. Mammalian sperm offer one compelling example. Long thought to be mere paternal DNA delivery vehicles, recent research with cutting-edge RNA sequencing technology revealed that, in mice, sperm cells harbored a variety of RNA molecules that transfer to oocytes during fertilization; experimental evidence suggested that this sperm-derived RNA is important for subsequent early embryonic development.[23,24,25] Even more striking, other recent research suggested that sperm RNA is capable transmitting environmentally induced (i.e., "acquired") traits to the next generation. Injection of sperm RNA into normal mouse zygotes generated offspring that recapitulate paternal traits, including mental stress and metabolic phenotypes such as obesity.[26,27]

Epigenetic inheritance also occurs by way of proteins. One well-studied example involves histones, proteins that form intimate interactions with DNA molecules. Enzymes called histone methyltransferases can add methyl groups to histone proteins. This enzymatic addition of methyl groups to histones impacts the transcription process; histone methylation is generally associated with transcriptional repression (reduced levels of RNA production), though activation can also result depending on which specific histone protein and region undergoes chemical modification.[28] The methylation status of histones and associated impacts on gene expression can be inherited from one generation to the next, and thus constitutes a form of epigenetic inheritance.

Other proteins are capable of both self-replicating and transmitting across generations. These forms of protein-based epigenetic inheritance do not rely on interactions with DNA and sometimes have dramatic biological impacts. Prions offer well-known examples; the ability of these proteins to transform from innocuous proteins in the cell into rapidly self-replicating molecules underpins devastating diseases such as Scrapie in sheep and goats, Creutzfeldt-Jakob disease (a.k.a. mad cow disease), and kuru.[29] Kuru, also known as "guria," is a rare, incurable, fatal neurodegenerative disorder that was formerly common among the Fore people of Papua New Guinea. People contracting kuru prions would suffer severe shakes and tremors; it is also known as the "laughing sickness" due to the pathologic bursts of laughter experienced by the infected.[30] It is now widely accepted that kuru was transmitted among members of the Fore tribe of Papua New Guinea via funerary cannibalism. Deceased family members were traditionally cooked and eaten, which was thought to help free the spirits of the dead. Women and children usually consumed the brain, the organ in which infectious prions were most concentrated, thus allowing for transmission of kuru. The disease was therefore more prevalent among women and children. In addition to transmission from one generation to the next through cannibalism, some self-replicating prions are capable of transmitting from one generation to the next through haploid gametes into zygotes.[31]

DNA does not hold a unique, special role as *the* molecule of heredity. RNA molecules can be inherited and influence traits in the next generation, as can proteins. Many of these knowledge advances in epigenetic inheritance arose over just the last few decades, due to brave research willing to challenge the Central Dogma and other long-held assumptions, such that the inheritance of environmentally derived traits is not possible. Conceptual barriers are breaking, and H_S1 remains supported.

* * *

Linear thinking often underpins conceptual models of inheritance. Mendel's pea plant experiments were designed to investigate the patterns by which traits were unidirectionally, or "vertically," inherited from parent to offspring. The subsequent development of genetic thinking continued to build upon this paradigm, especially as the details of DNA's role in inheritance became clear. A widely held assumption of universal vertical DNA inheritance, at least in animal and plant species, arose and took hold in the minds of many biologists. A strictly linear, vertical transgenerational inheritance of DNA paints a picture of heredity that is inconsistent with H_S1 (because DNA follows an invariant mode of transmission) as well as H_S3 (because DNA's transmission represents a linear rather than mutual chain of cause and effect, from parent to offspring).

The assumption of vertical inheritance underpinned the development of "tree of life" concepts, such as that developed by Charles Darwin and applied to his theory of evolution by natural selection. In Chapter IV of *On the Origin of Species*, Darwin wrote:

> The affinities of all the beings of the same class have sometimes been represented by a great tree. I believe this simile largely speaks the truth. The green and budding twigs may represent existing species; and those produced during each former year may represent the long succession of extinct species. At each period of growth all the growing twigs have tried to branch out on all sides, and to overtop and kill the surrounding twigs and branches, in the same manner as species and groups of species have tried to overmaster other species in the great battle for life.

"Tree of life" thinking led to phylogenetics, a branch of evolutionary-genetic science that investigates the evolution of life by inferring the relationships among organisms based on patterns of similarity in DNA sequences or other biological traits. The result of a phylogenetic analysis is a phylogenetic tree, which presents a hypothesis on the relationships among the organisms under study. Early molecular phylogenetic approaches of the 1980s and 1990s, usually based on DNA sequence comparisons, were highly successful, often providing important insights into the relationships of diverse animal, plant, and microbial lineages. For example, molecular phylogenies led to a proposed deep restructuring of the major forms of life, away from the "five kingdoms" model and the "eukaryote/prokaryote" dichotomy, and toward one defined by three broad "domains" of life: the archaea, bacteria, and eukaryotes.[32] During this time, most phylogenetic analyses relied on just one or a few regions of DNA that are widely conserved in different forms of life, such as ribosomal RNA genes or mitochondrial DNA.

However, as DNA sequencing became cheaper and phylogenetic analyses began incorporating multiple genes, it became clear that, sometimes, different genes told different stories. Discordant phylogenies resulting from analyses focused on different genes initially arose in analyses of the relationships among different bacteria, and between archaea and eukaryotes.[33,34] Increasing numbers of similar results accumulated as microbial evolutionists studied more and more genes from more and more microbes, with comparisons of complete genomes between different strains of the same bacterial species often showing that hundreds of genes likely evolved through horizontal transfer, between bacteria of different species.[35] These findings led to the unavoidable conclusion that horizontal exchange of DNA was a significant force in the evolution of bacterial genomes, which could not be ignored. For some this result was not too surprising, because there were many well-known biological mechanisms by which horizontal DNA inheritance can take place. In conjugation, small circular DNA molecules known as plasmids transfer from one bacterial cell to another through a tube-like pilus structure that bridges donor and acceptor cells. Phages are also able to transfer DNA from one bacterium to another, a process called transduction. A third example is transformation whereby bacteria simply uptake DNA from the environment. The ability of bacteria to evolve through horizontal gene transfer has many implications, such as the rapid spreading of antibiotic resistance genes and rapid responses of human gut microbiomes to changing gastrointestinal conditions.[36] The prominence of horizontal gene exchange in shaping microbial evolution led to a new paradigm of the "web of life," rather than the misleading (at least for bacteria) "tree of life" concept.[37]

Though less common than in bacteria, many examples of horizontal DNA transfer involving animals and other eukaryotic organisms have now been documented.[38] Many of the known cases involve exchanges between endosymbiotic bacteria, living inside the cells of other organisms, and their animal hosts. The widely studied *Wolbachia* lineage of endosymbiotic bacteria infects numerous invertebrate animal hosts and is transmitted through the female germline. *Wolbachia* can have both positive (e.g., increases in fecundity) and negative (e.g., killing of male progeny) effects on infected hosts. The first *Wolbachia*-to-host transfer of DNA was described in a bean beetle, *Callosoruchus chinensis*, where endosymbiont-originating DNA was discovered in one of the beetle's nuclear chromosomes.[39] Later, an entire ~1.4 million–base pair *Wolbachia* genome was found to be integrated into a nuclear chromosome of a fruit fly, *Drosophila ananassae*.[40] *Wolbachia* also infect nematodes; lateral symbiont-to-animal host transfer events have been detected in a nematode species, *Acanthochelonema viteae*, which infects rodents, and another nematode species, *Onchocerca flexuosa*, which infects deer.[41] Nematodes that parasitize plants have also acquired genes from bacteria through lateral transfer, and it is hypothesized that such exchanges enabled their free-living ancestors to exploit plant roots as a new ecological niche. The genome of *Meloidogyne incognita*, the "root-knot nematode," contains sixty genes of diverse bacterial origins that encode proteins required for the breakdown of plant cell walls.[42] Thus, horizontal DNA transfer is increasingly appreciated as a major evolutionary force across diverse forms of life. The varied vertical and horizontal modes of DNA transmission observed in diverse forms of life are consistent with H_s1.

Does DNA always transmit in a one-way fashion, from parent to offspring? The paradigm of Mendelian genetics suggests a one-way street for DNA transmission, with parental donor DNA flowing in one direction toward offspring recipients. But does DNA inheritance ever flow backward? It has been known since the late 1990s that fetal DNA makes up ~10 percent of the DNA molecules free-flowing in the plasma of pregnant women.[43] Though these DNA fragments circulating in arteries and veins are small in size (~150 base pair fragments on average), knowledge on the presence of fetal genetic material in maternal plasma has enabled the development of new noninvasive prenatal testing methods for genetic diseases.[44] Further, it is known that intact fetus-derived cells enter maternal bloodstreams during pregnancy and can persist for many years after pregnancy.[45] These fetal cells are then capable of colonizing different human tissues in the mother, leading to a condition termed microchimerism. One study that deployed a PCR assay for a Y-chromosome gene found that 63 percent of women who had given birth to sons harbored male microchimerism in their brains.[46] The son's Y chromosome–derived DNA was found in many regions of the mothers' brains. Microchimerism also has the potential to impact the mother's health: women with Alzheimer's disease were found to have a lower prevalence and concentration of microchimerism in their brains as compared to women without the neurologic disease. Thus, DNA does not follow a strict one-way inheritance avenue, from parent to offspring. Rather, it is capable of mutual, bidirectional inheritance, from parent to offspring and from offspring to mother. H_s3 remains supported.

Uniformity

A human body is composed of many trillions of constituent cells, all derived from a single zygote that previously derived from the union of sperm and egg. The DNA present inside the nuclei of all those cells resulted from numerous successive rounds of DNA replication events accompanying cell divisions. It is widely held assumption that the DNA is identical throughout the cells of the human body; this assumption is essential to numerous practical applications of DNA technology, such as the use of genetic identification in forensics. For example, matches between DNA samples collected in rape kits and DNA collected from those accused of sexual assault constitute critical pieces of evidence for ensuring that those who are guilty of such atrocities are accurately identified and punished accordingly. DNA information is also essential to freeing those wrongly accused of a crime, convicted before the genetic technology was available. But is DNA truly identical among all the trillions of cells in the human body? If yes, this would imbue the molecule with an unchanging nature within a human body that offers an aspect of "self" to humans, thus being inconsistent with H_s1 and H_s2.

It has long been known that DNA mutates across cell divisions within a human body, the process of somatic mutation. Such mutations are major underlying causes to the onset of sporadic cancers. However, it has been difficult to directly characterize the rate and spectrum of somatic mutation because such events are rare, and different somatic cell lineages in the human body are expected to have distinct, unique mutational histories through the process of development from zygote to adult. In 2017, a team of scientists from the Albert Einstein College of Medicine in New York developed a single-cell, whole-genome sequencing strategy to investigate the somatic mutation rate in humans and mice.[47] Focusing on small changes in DNA (such as a C:G base pair mutating to a T:A pair due to cytosine methylation), they discovered that the somatic mutation rate is nearly one hundred times higher when compared to the germline mutation rate (the rate of change from one generation to the next) in both species. Their results indicated that each cell in the human body likely harbored many hundreds of mutations distinguishing it from its trillions of somatic relatives in the body. However, since the human genome is more than three billion base pairs in size, only approximately one base pair in ten million is expected to mutate per cell division, leaving the remaining ~99.9999 percent of the genome sequence unchanged.

Another study coming from the Salk Institute in San Diego focused on genomic variation among neurons in human brains and offered a similar somatic mutation story, but it also revealed the prevalence of large structural genome changes among different neurons.[48] Some neurons were missing large chunks of DNA found in other cells, as well as duplications of large genome segments that existed as single copies in other cells. Thus, although ultimately every single one of the trillions of cells in a human body is expected to harbor numerous unique mutations that differentiate it from all other cells, the frequency of differences is sufficiently low that the somatic cells can safely be considered mostly identical. The *conventional* notion of somatic DNA uniformity, though potentially misleading and not true in strict senses, is essential to important practical uses in forensics and other applications of DNA biotechnologies. In an *ultimate* sense, however, H_S1 and H_S2 remain supported.

* * *

Medical interventions such as blood transfusions and transplantations can also lead to somatic variability within a human body, similar though larger in scale than the previously discussed concept of microchimer ism.[49,50] Recipients of organ transplants might be considered examples of "macrochimeras." Francisco Valera, a cofounder of the Dalai Lama's Mind and Life Institute dialogues, was the recipient of a liver transplant, and thus became a macrochimera. Valera was a Chilean biologist, neuroscientist, and philosopher who became a practitioner of Tibetan Buddhism in the 1970s through interactions with multiple teachers, including Keun-Tshen Goba (né Ezequiel Hernandez Urdaneta), Chogyam Trungpa Rinpoche, and Tulku Urgyen Rinpoche. In addition to catalyzing the Mind and Life dialogues with the Dalai Lama in 1987, Valera is well known for his introduction of autopoiesis to biology, in collaboration with his mentor Humberto Maturana.[51] The concept of autopoiesis was initially developed to describe the self-maintaining chemistry of living cells; subsequently, it was extended to the fields of cognition, systems theory, and sociology.

Valera contracted Hepatitis C and battled ever-worsening complications from the RNA virus infection in the latter stages of his life, including cirrhosis and then liver cancer. In 1998, he received a liver transplant in Paris and subsequently wrote a series of notes during his three remaining years of life, compiled into an essay titled "Intimate Distances—Fragments for a Phenomenology of Organ Transplantation."[52] Valera's writings about his transplant experience emphasized *anatman* (i.e., support H_S2) and illustrated the impact that Buddhism had on his biological thinking and sense of "self":

I emerged from the surgery with the liver of an unknown five days ago. My attention now turns to the two men as they speak. I follow their conversation and wait expectantly for words directed at me. It is a crucial moment: if the veins and arteries have not taken to their new place, my whole adventure comes to a halt. The graft, from their point of view, represents hardly anything more than a successful fixture. I am short of breath as I pick up the doctor's overheard telegraphic comments: Good portal circulation, no inflammation. . . . Abruptly he smiles at me and says: "Tout va bien!"

I've got a foreign liver inside me. Again the question: Which me? Foreign to what? We change all the cells and molecules of a liver every few weeks. It is new again, but not foreign. The foreignness is the unsettledness of the belonging with other organs in the ongoing definition that is an organism. In that sense my old liver was already foreign; it was gradually becoming alien as it ceased to function, corroded by cirrhosis, with no other than a suspended irrigation of islands of cells, which are then left to decay and wither away.

Sex

Charles Darwin published *The Descent of Man, and Selection in Relation to Sex* in 1871. In this book, Darwin expanded beyond his general theory of evolution by natural selection, focusing on the subjects of sexual selection and human morality and ethics. Darwin's extension of his theory to a range of questions traditionally discussed within philosophy, theology, and social sciences hardened the opposition of many religious individuals and communities to his evolutionary ideas. Darwin developed a novel approach to ethics, "exclusively from the side of natural history,"[53] rooted in his general theory of evolution and hypothesizing that human moral codes and cultural norms arose in human evolution as products of descent from nonhuman ancestors. The moral sense, according to Darwin, resulted from animal instinct deriving from biological descent, and particularly from the social dispositions resulting from natural selection. From this view, Darwin argued for an evolutionary origin of ethical foundations in humans, with the precursors of human ethical behavior found in the behavior of other animal species, particularly those with social organization. Darwin's *Descent of Man* was highly influential and contributed to the development of myriad novel disciplines and endeavors, including sociobiology, evolutionary ethics, evolutionary psychology, and the pseudoscience of eugenics.

More than a century after Darwin published his *Descent of Man*, another man from England wrote and published another book with the same title. Both authors bravely challenged the societies surrounding them with their works and words. Both books focused on the behavior of humans. The appearances and personal styles of these two English authors, however, were quite different. Whereas Charles Darwin was known for his long gray beard and reserved dark suits typical of late nineteenth-century Britain, the contemporary author Grayson Perry has at times in his life identified as a man but also frequently appeared and identified as a woman. Perry maintains a fluid sense of gender identity and has variably described his female alter ego, "Claire," as a nineteenth-century reforming monarch, a forty-something woman living a typical middle-class life in modern England, and as a rifle-toting Eastern European freedom fighter. Perry began dressing in women's clothes as a youth and was thrown out of his father's house at the age of fifteen as a consequence; he then went to live with his mother and stepfather. Perry is a renowned artist for utilizes a variety approaches and media, ranging from pottery to tapestries to clothing and fashion. In 2016, Perry published his *Descent of Man*,[54] a book focused on notions of modern masculinity that integrated biographical elements. In the book, Perry wrote:

> I sometimes watch the evening news on television and think all the world's problems can be boiled down to one thing: the behavior of people with a Y chromosome.

It is a commonly held genetic assumption that "males" have a single X chromosome and a single Y chromosome, and that "females" have two X chromosomes. This model of chromosome-based sex determination arose in the early 1900s, from the fruit fly experiments of Thomas Hunt Morgan and Lilian Vaughan Morgan at Columbia University in New York. This genetic paradigm resulted from the observed inheritance patterns of a gene found on the fly X chromosomes (*white*—when mutated, gives rise to flies with white instead of red eyes). The Morgans' fly crossing experiments provided the first evidence for sex chromosomes and are featured in virtually all undergraduate biology and genetics textbooks. These experiments, and variations on them, have been repeated countless times in high school and college-level teaching laboratories. The phenomenon of male color-blindness, previously discussed in Chapter 4, offers another popular example of sex chromosome-based inheritance. This chromosome paradigm for sex determination, however, appears to be at odds with the second Buddhist hypothesis. Does one's sexual identity strictly and invariably correlate with the status of one's sex chromosomes? Does the presence of a Y chromosome in a zygote, and subsequently in all adult somatic cells derived from it, present a rigid DNA-based determinant of inherent sexual "self" identity? Affirmative answers would be inconsistent with H_s2.

* * *

Sexual identity is sometimes determined by environmental rather than genetic factors. In a variety of lizard, snake, and turtle species, biological sex is determined by temperature, often during the middle third of embryonic development.[55] All crocodilian species have temperature-dependent sex determination, as does the single remaining lineage of the Sphenodontidae family of reptiles, the tuataras (*Sphenodon punctatus*). Tuataras are morphologically very similar to lizards and are only found in New Zealand. Like many other species native to these islands, tuatara populations severely declined after the arrival of humans and the other mammalian species that accompanied them. Until recently, tuataras were entirely absent from the two main islands of New Zealand and only inhabited thirty-two small offshore islands that have been kept free of mammals. However, a population of tuataras was established at the Zealandia wildlife sanctuary on the North Island in 2005, and a few years later a breeding nest was discovered.

In addition to threats from invasive mammalian species, tuataras face further extinction risk due to their temperature-based sex determination mechanism and the formidable force of global climate change. The biological sex of a tuatara hatchling depends on the temperature of the egg, with warmer eggs producing males and cooler eggs producing females. At 21°C, eggs have an equal chance of developing into a biological male or female. At 22°C, around 80 percent of eggs are likely to become male; at 20°C, approximately 80 percent of eggs become female. An international team of scientists from New Zealand, Australia, and the United States developed a mathematical modeling-based analysis of the potential fate of the tuatara under current predictions of climate warming by the mid-2080s.[56] If air temperature elevation occurs at the low end of the projections by this time (<1°C), tuatara sex ratios were projected to become increasingly male-biased. At the high end of the projections (4°C elevation), the model predicted 100 percent of tuatara nests to be 100 percent male by the 2080s. The occurrence of temperature-based sex determination in tuataras and other reptilian species has presented a biological puzzle investigated and debated by many evolutionary biologists,[57] one that is now made more urgent due to the special impacts of global climate change on species that reproduce this way.

The situation in species where sex is genetically determined, such as humans, is also not so simple. The binary biological concept of "XX = female and XY = male" is complicated by the reality that sometimes children are born with sex chromosome situations that deviate from this overly simplified and misleading model. A 1990 study, retrospective across a thirteen-year timespan in Denmark, revealed that 78 out of 34,910 children surveyed in this timespan harbored a sex chromosome status that deviated from the XX and XY "norms."[58] The most common situation observed was Klinefelter syndrome, where individuals have one Y chromosome paired with two or more X chromosomes. Such individuals are infertile and usually identify as biological males. Associated traits vary and are often subtle, not becoming evident until puberty. Triple X syndrome was the second most frequently observed case. Individuals with three X chromosomes typically identify as biological females and are fertile. Turner syndrome, where affected individuals typically have only one X chromosome instead of two,[59] was the third most common deviation. Those with Turner syndrome usually identify as female.

In Swyer syndrome, individuals born with XY chromosome states develop into individuals with anatomical features typically associated with biological females, and they usually gender-identify as women. People with Swyer syndrome have typical female external genitalia. The uterus and fallopian tubes appear as normal. Their gonads, however, are not functional; affected individuals have undeveloped clumps of tissue called streak gonads that sometimes become cancerous, so they are usually surgically removed early in life after diagnosis. How is it possible for an XY individual to develop into a female? There are many varied underlying genetic factors that can lead to this outcome. In ~15 percent of the cases, there is a mutation in the *SRY* gene on the Y chromosome; this gene encodes a transcription factor protein that plays a central role in male sexual development.[60] Though the genes on sex chromosomes do play important roles, sexual development requires many genes not located on X or Y. Swyer syndrome can also result from mutations in the *MAP3K1* gene, located on Chromosome 5. Mutations in this gene act early in human embryonic development, decreasing molecular signaling that leads to male development and increasing signaling pathways leading to female development.

Many people are born with any of several different kinds of variations in sex-related traits (e.g., chromosomes, gonads, hormones, genitalia) that do not fit into the typical binary conceptual framings of male and female bodies that are pervasive in human societies, especially in the West. Increasingly, some individuals with such traits that do not fit the "binary norm" use the term "intersex" to describe themselves. Estimating the frequency of intersex individuals is difficult because sometimes there are diverse varieties of intersex conditions, and anatomical traits are sometimes subtle and do not show up until later in life. Resources from medical centers and experts often quote frequencies around 0.05 percent (~1/2,000 births), though these are usually based on a restricted range of intersex conditions; a higher estimate of 1.7 percent (~1/60) was reported in a 2000 study from Australia that considered a broader range of conditions.[61]

Intersex people often suffer in distinctive ways. In addition to navigating life in societies where male/female binaries are presumed to be universal and inherent norms, intersex individuals are commonly subjected to genital "normalizing" surgeries early in their lives. Those in the intersex activism movement now refer to these interventions as "intersex genital mutilations." Oftentimes, these surgeries are performed based on the decisions of an intersex individual's parents and doctors, with a variety of physical-health and psychosocial rationales. Supporters of these surgical interventions claim the surgeries are often necessary to reduce risk of cancer in gonads, to improve the potential for fertility, and/or to close up exposed internal organs. Psychosocial justifications include making the child's appearance more consistent with their sex of rearing, to reduce the effects of atypical genitalia on psychosexual development and gender identity, and to alleviate parental distress over the atypical genital appearance of their child. All of these rationales are under heated debate. In 2011, Christiane Volling became the first intersex individual to successfully sue for damages associated with nonconsensual surgical intervention. In 2015, Malta became the first country to outlaw nonconsensual surgeries to modify sexual anatomy. Amnesty International and other human rights institutions have called for an end to all nonconsensual "normalizing" surgeries performed on intersex people.

Intersex activism swelled during the 2010s. Pidgeon Pagonis, an individual who self-identifies as gender nonbinary, was diagnosed with androgen insensitivity syndrome, a condition that impairs the masculinization of male genitalia during development, as a child. They were not informed of this condition, raised as a girl by their parents, told that they had ovarian cancer (which were really internal testes), and subjected to a series of surgeries to alter their genitalia.[62] Pagonis learned about intersex traits during their time in college, and then accessed their medical records and learned the truth about their condition. After graduating college, Pagonis joined the interACT advocacy organization and went on to become a central figure in the intersex activism movement, contributing to and creating video documentaries, essays, and social media campaigns to raise awareness of intersex issues. Pagonis was honored as an Obama White House Champion of Change in 2015. However, a 2018 Trump administration White House memo, reviewed and reported on by the *New York Times*,[63] indicated that the administration considered "narrowly defining gender as a biological, immutable condition determined by genitalia at birth."

Despite societal norms and ongoing objections in various political and religious spheres, there exists no chromosomal or other inherent, immutable biological basis that can invariably be used to assign people to concrete and binary "male" or "female" sexual identity bins. In other species, such as Nile crocodiles and tuataras, temperature is the central determinant of biological sex, not DNA. Our broader and growing knowledge of the complex interactions of environment, genetics, and society in shaping sexual identity across different forms of life strongly supports H_s2.

Race

James Watson, Nobel laureate and DNA double helix superstar, gave an interview to the *Sunday Times Magazine* during a 2007 book tour in the United Kingdom; he stated that he was "inherently gloomy about the prospect of Africa" because "all our social policies are based on the fact that their intelligence is the same as ours—whereas all the testing says not really." A few years later, Watson was at a meeting with two prominent geneticists of Jewish heritage—David Reich and Beth Shapiro—and asked them, "When are you guys going to figure out why it is that you Jews are so much smarter than everyone else?"[64] Watson asserted that Jews were high achievers because of genetic advantages conferred by thousands of years of natural selection to be scholars. In the same conversation, Watson claimed that East Asian students tended to be conformists due to natural selection for conformity in ancient Chinese society. No reliable scientific evidence, however, backs up any of Watson's racist claims.

Any analysis of "race" is complicated by the fact that this charged term has different meanings and is differentially applied, depending on the context. For example, many biological disciplines use the term "race" to specifically refer to a subgroup within a species that shares some shared geographic or biological trait affinities that distinguish those in one "race" from other "races" of the species. In *Meloidogyne incognita*, the previously discussed plant-parasitic nematode species that horizontally acquired bacterial DNA, scientists studying these nematodes have defined four different "race" categories that are defined by specific host plant species the different nematode races can infect.[65]

When applied to humans, the term "race" holds great power in the public arena and is accompanied by disagreements about what the term means and how it should be applied. On one end of the spectrum, some people conceptualize the human races as "real" natural biological categories reflecting shared genetics and ancestry. On the other end of the spectrum, some view the human races as pure social constructs, reflecting complicated and interacting effects of geography, religion, ethnicity, and so on. Those whose views align more with the former end of the spectrum, such as many white supremacists, often believe that underlying genetic differences strongly separate those of different races and explain other differences, such as in intelligence. Do clearly defined categories of humans, based on inherent and uniform genetic differences, exist that correspond to the different human races? Evidence of the affirmative would be inconsistent with H_S2.

* * *

In the nineteenth and early twentieth centuries, many Western biologists were obsessed with the concept of race. The then-prevailing evolutionary view was that the races represented long-separated subgroups of humans, with deep distinctions across numerous physical and psychological traits. A biological subdiscipline arose, "racial biology," that became a prominent topic of research and instruction at diverse institutions of higher education. Racial biology eventually contributed to the development of the discipline of eugenics, a pseudoscience deriving from genetics and evolution, which advocated for improving the human species through prescriptive reproductive practices.[66] Eugenics had a profound impact on American and European societies, resulting in the onset of government policies forcing involuntary sterilization surgeries on thousands of individuals categorized as "feeble-minded," "sexual deviants," and other traits that were erroneously assumed to result entirely from the genes.

Eugenic pseudoscience became a staple of higher education in the early twentieth century. During this time, shortly after the rediscovery of Mendel's results but before the dawn of DNA, European and American scientists obsessively focused on empirically and 'objectively' defining differences between the races through various comparisons of skulls. A strong desire to demonstrate real differences between the races motivated their hypotheses, particularly in traits presumably related to intelligence. In *The Mismeasure of Man*,[67] the late great Stephen Jay Gould systematically dissected and debunked decades of scientific research activity that, at some time or another, claimed to offer support for race-based and other differences in intelligence. These critiques have developed into classic examples of bad science, conducted by racist scientists. For example, Gould dove into the research of Samuel George Morton, now known as the "Father of Scientific Racism." Morton was a physician and "natural scientist" living and working in Philadelphia during the early nineteenth century. Morton was also a tomb raider. During his life he traveled to collect skulls from Egyptian tombs, Native American burial grounds, and other places around the globe; he ended up with more than one thousand skulls in his collection. Morton measured brain volume in the skulls and predictably found that those of the "white" race were larger than those from other races. Morton performed his early experiments by filling skulls with pepper seeds (which could be packed down); his later work used lead shot to fill the skulls instead. Some skulls were measured with both methods and revealed a pattern of inconsistency suggesting that Morton tended to pack "Caucasian" skulls more tightly with mustard seeds (leading to inflated brain volume estimates) and "negro" skulls more loosely (leading to brain volume underestimates). Gould identified many specific problems with Morton's methodologies and analyses:

- *"Favorable inconsistencies and shifting criteria."* Morton often chose to include or delete large subsamples in order to match group averages with his prior expectations.
- *"Subjectivity directed toward prior prejudice."* Morton's seed measures were more imprecise than those with lead shot. Comparing numbers from skulls examined by both methods offers a window into Morton's

prior prejudice. Blacks fared poorest and whites best when results could be biased toward expected results. (In other words, values for Black people were always downwardly biased whereas values for white people were upwardly biased in seed-based measurements).

- *"Procedural omissions that seem obvious."* Morton never corrected for overall body size, differences in sexes, and so on. Many of his "negro" measurements were based on women only, and many of the "Caucasian" measurements came from men-only samplings.
- *"Miscalculations and convenient omissions."* Morton made dozens of basic math errors. When Gould redid all of the Morton's calculations, he discovered that every single one of Morton's errors was in favor of his hypothesis. A large Chinese skull was omitted from Morton's analysis (for no apparent reason), along with other similar mysterious omissions.

In *The Mismeasure of Man*, Gould also highlighted the scientific work of Robert Bennett Bean, a Virginia physician performing research on racial differences during the early twentieth century. Rather than focusing on skull volume, Bean studied the postmortem brains of American white and Black people by measuring the relative lengths of two parts of the corpus callosum: the genu and the splenum. The genu was at the front of the corpus callosum and believed to be the place of higher mental functioning; the splenum was at the rear of the corpus callosum and thought to be the locale of lower and more basic mental functions. Bean hypothesized that the races could be separated based on the relative lengths of the genu and the splenum. When Bean completed his measurements on dozens of human brains, the results supported his hypothesis. Brains coming from white people had high genu-to-splenum length ratios, whereas those from Black people had distinctly lower ratios. However, Bean's mentor at Johns Hopkins University, Franklin Mall, repeated the measurements for himself, suspicious that Bean's results might be too good to be true. Mall got very different results. The mentor's measurements revealed no differences whatsoever in the genu/splenum length ratios between white and Black people. There was one key methodological difference between the two sets of measurements: Robert Bennett Bean knew the racial identity of the brains *before* he measured them, whereas Franklin Mall performed the measurements without prior knowledge of their races. Bean's scientific misadventure now serves as a classic cautionary tale, frequently shared in biology classrooms and medical school lecture halls, of past racial bias and the importance of blind trials and randomization in experimental design.

Before moving on, the influence of racial and eugenic attitudes in the world of Buddhism warrants discussion. As detailed by Donald Lopez in *Buddhism and Science: A Guide for the Perplexed*, the writings of Anagarika Dharmapala, the champion of Buddhism central to its postcolonial revival in the early 1900s, showed evidence of the prevailing racial and eugenic attitudes in the Western world. Dharmapala often emphasized the "Aryan" birth of the Buddha and noble bloodlines of the Sinhalese people of Sri Lanka in his writings, often contrasting them with those of less noble "Semitic" and other ancestries. In "Evolution from the Standpoint of Buddhism," Dharmapala criticized Western religions through the use of racial and eugenic assertions:

> Where is the justice in the doctrine [of Christianity, Judaism] when we see so many millions of insane, ignorant, feebleminded, deaf, dumb, crippled, blind, suffering from incurable diseases on this earth, and the semi-savage half-animal people of Africa? Are they all to go after death to an eternal hell-fire?

A sympathetic, apologist attitude might downplay such expressions written by Dharmapala. The Anagarika's privileged interactions with Western white scholars, something not common among most Sri Lankans at the time, shaped his view of the races and were part of his broader mission to align Buddhism with the modern scientific views of the time (contrasted with Christianity's inability to do so). However, it is important for supporters of Buddhism, and the Buddhism and science dialogue, to recognize that its central historical figures were often strongly shaped by Western influences of all manner, including the pseudosciences of racial biology and eugenics.

* * *

Has evidence for genetic differences between the human races arisen in the modern, DNA-informed era? Molecular approaches to characterizing human race relationships began in the early 1970s, with Harvard evolutionary geneticist Richard Lewontin publishing a pioneering study of natural variation in human blood proteins.[68] He arranged the analyzed human populations into seven races categories based on his findings: Africans, Australians, East Asians, Native Americans, Oceanians, South Asians, and

West Eurasians. Lewontin found that ~85 percent of the overall blood protein variation could be accounted for by differences *within* populations and his race categories, and only ~15 percent of the variation was distributed across them. To the extent that there was variation among humans, Lewontin concluded, most of it was because of "differences between individuals" rather than differences between races. This was the first molecular domino leading to a cascading consensus that, among human populations, there exist no between-group genetic differences big enough to justify the concept of a "biological race." Instead, with additional input arising from the social sciences, human races became instead conceptualized as a "social construct," a fluid and relative way of categorizing people that changes over time and varies from one part of the world to another.

Early DNA-based studies supported Lewontin's initial findings based on protein-level variation, and the onset of genome-wide approaches to investigating the genetic structure of human populations around the world revealed striking patterns of migration and mixing across human evolutionary history to further argue against the concept of a biological race. Svante Pääbo, director of the Max Planck Institute for Evolutionary Anthropology in Germany who is a leading expert on ancient DNA sequencing and hominin evolution, gave a TED Talk in 2011 titled "DNA Clues to Our Inner Neanderthal."[69] The talk focused on the sources of genomic variation in the global human population. Pääbo highlighted the extensive genetic diversity of people living inside Africa, widely cited evidence of *Homo sapiens'* origins on the continent, and the very recent migration of humans out of Africa to colonize the rest of the world, ~50,000 to 100,000 years ago. Pääbo next shared the results of an analysis of genome-wide patterns of DNA variation across 185 people living in Africa along with 184 people from Eurasia (Europe and China); powerful new DNA sequencing technologies enabled scientists to identify 38,877,749 variable DNA positions among those 369 people whose genomes were studied. The question was posed, "Can we find any absolute differences between Africans and non-Africans?" Pääbo emphasized that the population genomic data offered more than thirty-eight million opportunities to answer that question. Did any of the genetic variants demonstrate 100 percent differentiation between African and non-Africans? The answer was no. With relaxed criteria, allowing for just 95 percent differentiation between Africans and non-Africans, then only twelve out of the millions of variant sites show this pattern. Pääbo went further to reveal that Neanderthal DNA was present inside the genomes of modern-day humans of Eurasian descent,[70] and he concluded the TED Talk with the exclamation "We have always been mixing!"

The high degree of mixing and recent origins of modern human populations paint a picture of highly distributed worldwide genetic variation that is inconsistent with the "biological race" concepts that preoccupied the minds of early twentieth-century biologists and continues to prevail in many modern minds today despite all scientific evidence pointing to the contrary. The social construct theory of race remains widely held as the proper framework among many circles in society. Caution, however, is warranted in taking this view too far. In the same 2018 *New York Times* article where he shared his troubling interaction with James Watson, geneticist David Reich highlighted the potential utility of relying on "race" categories to identify genetic variants underlying prostate cancer. African Americans have a ~70 percent higher chance of prostate cancer, as compared to European Americans. Reich wrote:

> Self-identified African-Americans turn out to derive, on average, about 80 percent of their genetic ancestry from enslaved Africans brought to America between the 16th and 19th centuries. My colleagues and I searched, in 1,597 African-American men with prostate cancer, for locations in the genome where the fraction of genes contributed by West African ancestors was larger than it was elsewhere in the genome. In 2006, we found exactly what we were looking for: a location in the genome with about 2.8 percent more African ancestry than the average.
>
> When we looked in more detail, we found that this region contained at least seven independent risk factors for prostate cancer, all more common in West Africans. Our findings could fully account for the higher rate of prostate cancer in African-Americans than in European-Americans. We could conclude this because African-Americans who happen to have entirely European ancestry in this small section of their genomes had about the same risk for prostate cancer as random Europeans.

Thus, genetic whispers of our past geographic isolation remain in scattered parts of the human genome and are important not to overlook. However, the broader story of genomic mixing and greater intrapopulation genetic variation compared to interpopulation variation dominates the overall pattern of human DNA variation, and it is inconsistent with any meaningful notion of "biological races" in the human species. H_s2 remains supported.

Review

Four identity-associated facets of DNA—its roles in inheritance, presumed uniformity throughout the human body, role in sexual identity, and association with racial identity—were investigated in this chapter. These four facets built upon three features of DNA examined in the previous chapter (biophysical aspects, coding functions, and mutation process). Multiple efforts to falsify the three specific Buddhist hypotheses along the way all fell short. H_s1 predicted impermanence for all features of DNA. The misconception that DNA is the sole vehicle for genetic inheritance was broken down through investigations of RNA viruses, sperm RNAs, and prions. Strict vertical inheritance of DNA was nullified by diverse examples of horizontal DNA inheritance across diverse forms of life. High rates of somatic DNA mutation and associated natural variation across different cells that make up a human body supported H_s1. These observations of DNA nonuniformity within a person also offered support to H_s2, that DNA lacks characteristics associated with "self" nature. Further support for this second hypothesis arose from the erasing of the common misconception that sexual identity is strictly sex chromosome-based, and that racial identity has little to no genetic basis. H_s3 predicted that DNA would function in a mutual cause-and-effect framework, as opposed to the linear cause-and-effect conceptual models typical of Western thinking. The pattern of mutual inheritance associated with pregnancy, with DNA transmitting (as expected) from parent to offspring but also (surprisingly) from offspring to parent, offered a striking example supporting H_s3. As in the previous chapter, no "first cause" or "single cause" associated with DNA was uncovered in the journey examining this molecule's role in identity formation.

The thorough and in-depth efforts to falsify the three Buddhist hypotheses presented over the last two chapters, focusing on seven different facets of DNA, have fallen short. As with any hypothesis-driven investigation, these efforts offer no sort of "proof" of the Buddhist hypotheses, and the possibility will perpetually remain that other information or some investigator more insightful than I might identify avenues to their falsification. The approach was also limited, focusing on DNA, and as a consequence left untouched other features of the Buddha's teachings (e.g., *duhkha*, human rebirth) that were refractory to a DNA-centric inquiry. Despite these limitations and caveats, it is remarkable that a century of research and debate on the nature of DNA, the most iconic biological molecule in the world, has provided strong support for three propositions set forth by the Buddha more than two millennia ago, at least in the opinion of this scientific mind.

* * *

Before moving on, it is important to take one last look back and perform a "pseudoscience check." After all, the hypotheses analyzed derived from the teachings of the Buddha and Buddhism, and some readers might be worried that the author might have been brainwashed or is working from some other form of false logic or illegitimate reasoning. The retrospective pseudoscience check will be performed using Derkson's "seven sins of pseudoscience" framework presented in Chapter 6.

Derkson's first sin of pseudoscience, *"dearth of decent evidence"* does not seem to apply here because the sources of evidence used in this inquiry derived from ninety-two scientific, peer-reviewed articles and books focusing on diverse facets of biology and genetics, published across the last ~150 years. If the scientific literature is to be trusted, then this analysis based upon it should also be trusted. The second sin of pseudoscience, *"unfounded immunizations,"* is not relevant because no newly generated empirical data were handled or manipulated. Only the results and interpretations of the authors of the ninety-two papers, along with modern consensus knowledge about biological processes in the scientific realm, were relied upon. In fact, I purposely avoided referring to any of the scientific research articles that I have authored over my two decades as a publishing evolutionary geneticist to guard against the second sin. The third sin, *"ur-temptation of spectacular coincidences,"* also does not apply. Although coincidence has been a theme among scientific critics of Buddhism's evidence presented for human rebirth, the present analysis avoided this topic entirely and instead focused only on the specific teachings of the Buddha that were suitable for a hypothesis-based investigation. The fourth sin, *"magical methods,"* was avoided by limiting the analysis to peer-reviewed scientific literature that employed typical and reliable genetic experimentation and analysis approaches used by biologists all over the world. The fifth sin, *"insight of the initiate,"* is not relevant; there is nothing particularly special about the approach and evidence presented in the analysis. In fact, the investigation followed a conservative and stuffy Popperian hypothesis-testing strategy and came to general conclusions shared by others scientific minds such as Denis Noble and David Barash. The sixth sin, *"the all explaining theory,"* does not hold because the analysis was purposefully limited to specific sets of hypotheses relevant to a singular biological study subject, DNA. Preliminary hypotheses not appropriate for the approach deployed, related to the subjective human-experience phenomena of *duhkha* and rebirth, were discarded during hypothesis development in Chapter 6. As for the seventh and final sin, *"uncritical and excessive pretension,"* I have done my best to present the evidence in an even-handed fashion, purposefully seeking out and digging into features associated with DNA (e.g., the "universal" genetic code, presumed exclusive role in inheritance) that, if true, would falsify one or more of the hypotheses analyzed. Thus, to the best of my ability, features expected to accompany pseudoscientific ventures were avoided in the inquiry presented here. The investigation now concludes with full support for the three Buddhist hypotheses.

9

Bodhi

The analytical tour through scientific understanding of DNA supported the three Buddhist hypotheses. So what? Some Buddhist thinkers might characterize the entire endeavor as totally unnecessary. After all, the concepts of *anitya*, *anatman*, and *pratityasamutpada* have already undergone extensive analysis by numerous Buddhist minds over the last two and a half millennia. Many from the world of Buddhism would claim that the best evidence for the validity of these hypotheses comes from direct intrapersonal meditative experience, not from the analytical excursions of a busy science mind. Others might see the preceding chapters' lengthy tour through DNA, as viewed from Buddhist lenses, as a big distraction from Buddhism's central mission of extinguishing *duhkha*. Was the entire endeavor merely another example of the Parable of the Poison Arrow? Am I like that arrow-struck man who absurdly asked numerous questions about the details surrounding his life-threatening wound, demanding answers before he allows the doctor to pull the arrow out? Other modern-day evolution-focused minds who self-identify as "quasi-Buddhists" or "Buddhist wannabes," such as David Barash and Denis Noble, have already detailed some of the harmonies between Buddhist and biological thinking. Was any of this necessary?

For me and the *duhkha* I experience and see, the DNA-guided analysis of the three Buddhist hypotheses was absolutely necessary. I suspect the same for other science minds. First, despite the extensive literature and growing prominence of the "Buddhism and science" scene, including works coming from widely respected scientists deeply embedded in Western academic systems as discussed in Chapter 5, the vast majority of biologists and other scientists out there continue to be mostly unaware and/or unappreciative of the potential for productive synergy at this intersection. Effective communication and rationalization of the parallels and complementarities of biology and Buddhism to science-guided and other Western minds require the space to fully describe and unpack many unfamiliar concepts and ideas. Most of my close colleagues in the world of biology are supportive of me pursuing this unusual work, but also sometimes bring forth puzzled looks and skepticism when I try to integrate some Buddhist ideas into "regular" scientific conversations. I realized early on in this initiative that it would take a book, and direct application of the scientific method, to fully make my points.

The formal scientific support for the three Buddhist hypotheses provides deep-down validation of core teachings of the Buddha, at least for this scientist. The analysis was purposefully designed and executed using a stuffy, conservative hypo-deductive approach held sacred by many science minds. There was an explicit double-check for pseudoscience at the end of the effort, which came up empty. The outcome of this DNA-based analysis, and similar ones that I have done on the side through the years focusing on other biological entities (trees, worms, mitochondria, viruses, etc.) came up with the same basic result. This has had a transformational impact on me—how I view my "self," and how I view my activities and ethical responsibilities in science, and beyond.

What moral and ethical compasses guide a scientist's decisions and actions? Why might one choose to create a vaccine against a pandemic-causing coronavirus rather than engineer new viruses as biological weapons? Many scientists might answer with something along the lines of "for the sake of humanity," "for the common good," or "because it's what is right." Diffuse answers such as these might satisfy some, but others might find fault due to a lack of grounding in explicit moral and ethical frameworks. There are many ongoing science activities in the world right now that could easily be described as in line with the Buddha's mission of eliminating human suffering and other forms of *duhkha*: developing new medicines, reducing anthropogenic impacts on the environment, and developing new efficient green energies offer a few examples. Other ongoing science efforts purposefully contribute to human suffering. Some scientists weaponize bacteria and viruses. Some scientists make weapons more deadly. Some scientists cause intentional pain and suffering to animals. We scientists need strong ethical frameworks to help guide our decisions, but this is not something that many practicing scientists of today often talk about, at least in my experiences in my science circles. That needs to change.

Here I propose the Buddha's teachings as an ethical foundation for scientists—to guide the decisions we make and the science we do. This was the third and final book objective initially described in Chapter 1. Four qualities of Buddhist wisdom—selflessness, detachment, awareness, and compassion—are proposed to guide and motivate this way of doing science.

I call this Buddhist wisdom-based approach 'Bodhi science'. The Bodhi science strategy also serves as an immunizing agent to help prevent science transforming into pseudoscience.

Selflessness

Selfishness and ego drive have become normalized in science. This is due, at least in part, to some famous science personalities and their well-known egocentric attitudes. J. Craig Venter, the yacht-faring sequencer of DNA in humans and ocean microbes, credited his ego as the driving force responsible for his scientific achievements.[1] Another famous and controversial white-male science ego, Jim Watson, is also renowned for his strong sense of self:[2]

I'm the most accomplished person living on earth. I'm the top dog!

Scientists' egos, our false senses of "self," can be significant shapers of the course of science that are often overlooked. The illusory and imaginary "selves" of science compete with one another. We race to be the first to discover, the first to publish, the first to sequence the genome. We want to find *the* explanation for this or that, *the* gene for this or that. We self-apparitions of science want our names on the research articles, and we want our names to be first. We want the credit. We want the attention and respect of our science peers and of the world. We thirst for prestige. We want our names to be remembered, to be put on buildings. We hungry ghosts of science are never satisfied or content with our past accomplishments; we constantly crave more and more and more.

The Buddhist principle of *anatman*, supported in the investigation of the preceding chapters and across twenty-five centuries of Buddhist analysis, poses that there is ultimately no "self" there at all. Thus, any activity—scientific or otherwise—resulting from 'self' motivation or ego drive suffers from a foundational error of ignorance, according to the Buddha's teachings. Although a self-serving competitive attitude might bring temporary senses of superiority-based satisfaction with every little scientific conquest, the hunger for more achievements and more attention always returns. This is the *duhkha* of science.

On top of the influence of superstar science personas and the resultant normalization of self-glamorizing attitudes in science, the nature of the scientific method and meta-methods discussed in Chapter 6 poses additional challenges. The ideal of the "neutral and impartial observer" role of a scientist in the scientific process, tracing its roots back to Aristotle, leads to an erroneous conceptual framework where scientists are outside of the system under study. This flawed sense of separation, that the investigator is a negligible variable that can be ignored, is problematic. It dresses ego-motivated hypotheses, experiments, and interpretations in a deceptive cloak of false neutrality. This cape of illusion hides the truth of underlying ego motivations from others and from the scientists themselves. This is more *duhkha* of science.

In the Buddha's principle of *pratityasamutpada*, his inescapable mutual cause-and-effect framework, all interacting *dharmas* (phenomena) influence one another's existence. They are simultaneous causes and effects of one another. This applies to the scientific experiment and the scientist conducting the study. Operating in ignorance of this truth leads scientists to falsely assume that our ego desires and emotions have no impact on the science we produce, leading to the potential for subjective biases in our results about which we remain unaware. The nineteenth- and twentieth-century scientific racists, discussed in Chapter 8, offer examples of this process playing out. Dr. Morton tightly packed mustard seeds into the skulls of white subjects and loosely packed the skulls of Black subjects, resulting in inflated brain volume estimates for the former and downwardly biased estimates for the latter. Dr. Bean's prior knowledge of the racial identity associated with his brain measurements biased his results in favor of showing differences between the races. These scientific racists really wanted there to be brain volume differences between white people and Black people, and their numbers and measurements reflected this prior prejudice. In the light of *pratityasamutpada*, it can be seen that these scientists' racist attitudes influenced their science, and their racism-influenced results reflexively influenced the scientists by reinforcing preexisting racist attitudes in their minds. This trend continues today with the likes of Jim Watson and white supremacists spouting unsubstantiated claims of genetically determined IQ differences between white and Black people. All scientists, then and now, are subject to *pratityasamutpada*. The science that we decide to do, the hypotheses we decide to pursue, are in part a function of our egos and senses of self.

* * *

What can we do? What preventative steps can one take to avoid falling into ego traps and pursuing delusional "self"-motivated hypotheses? What is the path to selfless science? Here, as part of the Bodhi science strategy, I propose some mindfulness-based actions to help prevent selfish science, and conduct scientific research in a wiser fashion.

A Bodhi scientist acknowledges that they always harbor the potential for ego motivation and proactively seeks out its manifestations—the ways in which it "shows up." More broadly, scientists should maintain constant awareness that they are part of the study system, not separate from it. A return to the five *skandhas* ("aggregates") framework, the Buddhist mechanism of subject-object interaction and interdependence presented in Chapter 2, will prove helpful here. Recall that the first *skandha* is matter, followed by the second *skandha* of sensation, where there is initial contact made between object and subject. This leads to initial mental recognition through perception, the third *skandha*, which is followed by the construction of mental formations in the thinker's mind, the fourth *skandha*. The fifth *skandha*, consciousness, results from the countless simultaneous and temporally dynamic ongoing interactions described through the previous four *skandhas*. This five *skandhas* context provides an explicit descriptive framework for understanding the intimate interactive relationship between object and subject, how a person engages with the universe around them. No one escape this according to the Buddha's teachings, scientists included.

Many scientists, however, do not typically think about themselves and their relationships to objects of study in this five *skandhas* fashion. Instead, as previously discussed, we often prefer to think of ourselves as completely separated from our study targets. Scientists usually prefer to view their roles as being impartial observers with little to no impact on the object of study, and with the object of study having little to no impact on them. According to the Buddhist tenet of *pratityasamutpada*, operating through the five *skandhas*, this outlook is unwise and inconsistent with reality.

Instead of ignoring ego influences, or pretending like the impacts of false senses of "self" are not there, a wiser path is to directly acknowledge and confront this reality. In practice, this can be done through mindfulness-based reflective analysis. To begin, consider the *Satipatthana Sutta* ("The Foundations of Mindfulness" *Sutta*) of the Pali Canon, where the Buddha offered mindfulness advice on achieving full awareness to a group of his renunciate followers staying with him in the town of Kammasadhamma:

> Again, *bhikkus*, a *bhikku* is one who acts in full awareness when going forward and returning; who acts in full awareness when looking ahead and looking away; who acts in full awareness when flexing and extending his limbs; who acts in full awareness when wearing his robes and wearing his outer robe and bowl; who acts in full awareness when eating, drinking, consuming food, and tasting; who acts in full awareness when defecating and urinating; who acts on full awareness when walking, standing, sitting, falling asleep, waking up, talking, and keeping silent. (MN 10.8)

The Buddha was renowned for his ability to adapt his teachings to different audiences. To a group of Bodhi scientists, those wishing to follow Buddhist teachings to guide their scientific endeavors, the Buddha might adapt the preceding teaching given at Kammasadhamma, and instruct something like:

> A Bodhi scientist is one who acts in full awareness when asking questions, developing hypotheses, and designing experiments; who acts in full awareness when observing and collecting data; who acts in full awareness when organizing and analyzing data; who acts in full awareness when interpreting results and sharing them with other people.

We scientists should be willing and able to ask ourselves the hard, critical questions—to be aware and perform some self-reflective analysis before, during, and after scientific experiments. Why am I doing this? Do I have a vested interest in one outcome or the other? Have I somehow unwittingly set this experiment up to "succeed" or to "fail"? Do I hide some results? Do I exaggerate other results? Do I inflate the importance of my science activities when talking with others? Is this about finding the truth, or about others paying attention to me and my "self"?

Understanding *anatman*, that scientists and all humans lack a "self" nature, is critical to performing wise science. The Buddha's teachings on the five *skhandas*, *pratityasamutpada*, and mindfulness provide instruction on proper understanding of how scientists influence the science that they produce and how those science products influence them. A Bodhi scientist pursues inquiry with constant mindful awareness of the potential for ego influences, for the influence of selfish motivations underlying their decisions and activities. A Bodhi scientist does their science without clinging to a sense of "self." A Bodhi scientist is selfless.

Detachment

Scientists form attachments to hypotheses. We cling to our ideas, to our viewpoints. This is more science *duhkha*. Science's value of neutrality idealizes an archetypal investigator that is completely dispassionate about the support, or lack thereof, for a given hypothesis that they might have proposed. The Popperian ideal goes further, suggesting that scientists should not only be neutral regarding their hypotheses but rather proactively attack them. Despite all of this, in my experience, we practicing scientists frequently form strong attachments to our hypotheses and can become quite defensive and stubborn when our special ideas are challenged. Linus Pauling, the protein lover discussed in Chapter 7, preferred hypotheses involving proteins as the carrier of genetic information even after evidence accumulated that the genes were more likely in the nucleic acids. A second example derives from the field of theoretical population genetics. Moto Kimura, a self-taught mathematician and geneticist from twentieth-century Japan, developed the now-famous (at least among population geneticists) neutral theory of molecular evolution.[3] This theory presented a neutral, null hypothesis for the evolution of genes and proteins based on the nonselective evolutionary forces of mutation and genetic drift. This theory was at odds with the natural selection-minded evolutionists that prevailed then and (for the most part) today. Dr. Kimura became quite obsessed with his neutral hypothesis, as evidenced from an obituary written by his mentor at the University of Wisconsin, James Crow:[4]

> The neutral theory appeared in 1968, when Moto was 44 years old. He devoted the rest of his life to advocating and defending it. With single-minded determination he wrote one article after another and a widely read book (Kimura, 1983). The theory became an obsession. Everything he read was viewed from the standpoint of its bearing on his theory, and he selected evidence from many sources. Criticisms were regarded not as challenges that might lead to refining or broadening the theory, but as personal affronts.

The second Noble Truth of Buddhism identifies craving and clinging as the causes of *duhkha*. From a Buddhist lens, Kimura's preoccupations with his precious neutral hypothesis were key causative agents of the Japanese geneticist's science *duhkha*. Kimura clung to the idea that his neutral theory was the final answer, the ultimate evolutionary hypothesis. The theory was and remains highly influential in evolutionary biology circles; most evolutionary geneticists working today would probably agree that Kimura's hypothesis is very useful but also, at the very least, has limitations.

Some argue that the entire hypothesis approach imposes serious limitations on science. The Popperian hypothetico-deductive model, introduced in Chapter 6 and applied in Chapters 7 and 8, remains a revered ideal for many science minds. It is also a widely practiced approach among today's working scientists. Others, however, argue that hypotheses also bring "blinders" that limit science. For example, two contemporary computational biologists—Itai Yanai from New York University and Martin Lercher from Heinrich Heine University in Germany—wrote that hypotheses become liabilities when the mind of an investigator narrows their focus on a specific issue or question, or on a specific data set.[5] To make their point, Yanai and Lercher compared the hypothesis-guided scientific process to the famous "gorilla" selective attention experiment: study subjects were told to watch a video clip of students passing a basketball back and forth to one another. There were two teams in action wearing different colored shirts, and subjects were told to count the number of passes made by one of the teams. About halfway through, a person dressed up as a gorilla entered the scene in the video. The gorilla paused in the center, pounded its chest with its fists, and then exited to the opposite side of the frame. Surprisingly, half of study subjects in this experiment missed the gorilla entirely, because they were keenly focused on counting passes. In parallel control studies where video observers were not given the prior assignment of counting passes, the gorilla was detected by almost everyone. Yanai and Lercher argued that the same basic problem accompanies hypothesis-guided science—who knows what gorillas we are missing because we are busy counting things as directed by our hypotheses?

Candrakirti, a Madhyamaka Buddhist philosopher living in seventh-century India, would also likely find fault with science's hypothesis approach. Candrakirti was a key figure in the Prasangika school of Madhyamaka philosophy which rose to great prominence, especially in Tibet where the school's teachings underpin the Geluk tradition of the Dalai Lamas. Candrakirti's texts expounded upon the written works of Nagarjuna, the second-century Buddhist philosopher and originator of Madhyamaka philosophy. In works such as the *Madhyamakavatara* (Entering the Middle Way),[6] Candrakirti criticized all forms of statement-based approaches to logic, such as the autonomous syllogisms that were used by other Buddhist schools. For example, he attacked the foundational propositions used for "grounding" epistemology (theory of knowledge, way of knowing) by those in the Dignaga school that developed in India during the two centuries prior to Candrakirti. As expressed by contemporary comparative Buddhist philosopher Dan Arnold:[7]

> Candrakirti goes on to argue that if the questions and phenomena under consideration are properly understood, then the epistemologist's question ought not even to arise. Indeed, Candrakirti will argue that the epistemologist's imagined requirements are themselves evidence of precisely what is the problem to be overcome.

For Candrakirti and other Madhyamaka philosophers of the Prasangika tradition, the mere act of articulating a proposition, claim, or hypothesis for subsequent analysis brings forth accompanying elements of ignorance. Candrakirti also considered seeking universal explanations, Derkson's sixth sin of pseudoscience, as an especially problematic path leading to *duhkha*. We scientists often become attached to our own ideas, our own words, our own hypotheses. We cling to the idea of being right, and we avoid opening ourselves up enough to other perspectives or the pursuit of new lines of inquiry. We can't let go. This is science *duhkha*.

* * *

Scientists cling to categories. Biologists, for example, form attachments with their beloved biological units of choice: species, populations, organisms, cells, organelles, molecules. Linus Pauling cherished his proteins. Jim Watson and Craig Venter became attached to their DNA. Ornithologists obsess over their birds. Entomologists fixate on their insects. Conservation biologists cling to their endangered species of choice and become depressed about their species' pending departure from the earth's ecosystems. I have formed attachments to nematodes, anemones, banana slugs, Bodhi trees, and many other life forms during my years as a biologist. We scientists are susceptible to seeing individual members of these categories as fixed entities, and then proceeding to count and analyze these units to gain insights into biological processes, all the while forgetting (or not knowing in the first place) about the reality that there is no inherent existence in these categories we count. What is it that we are counting?

We scientists cling to our numbers. We value the apparent anonymity and objectivity that accompany quantitative approaches. We are susceptible to the belief that "numbers don't lie," and that our numerical data are free from personal biases and motives. Drs. Samuel George Morton and Robert Bennet Bean, the nineteenth- and twentieth-century scientific racists discussed in Chapter 8, believed in the inherent truth and power of their numbers. They were both misled by their egos and biases. We scientists see 1 as 1, and 0 as 0. We see 1 and 0 as completely separate and mutually exclusive possibilities. A Buddhist thinker, however, might see this situation quite differently. For example, Vietnamese monk Thich Nhat Hanh's interbeing concept[8] might instead suggest that 1 and 0 "inter-are": 1 is there only because 0 is there, and 0 is there only because 1 is there.

Scientists cling to science. We see science as the singular portal to the truth; many nonscientists in society today also revere science in the same way. This view brings us a sense of security and a sense of pride. However, as with science's hypotheses, myopic limitations accompany singular visions of science as the exclusive and privileged path to understanding the nature of things. In particular, science's reliance on the meta-methodological ideals of objectivity and quantifiability constitutes a significant constraint that is often overlooked.

According to the Buddha's teachings, objectivity is an impossibility. Those that claim to operate in the realm of pure objectivity suffer from delusions of false neutrality. Buddhism's mutual cause-and-effect framework, *pratityasamutpada*, posits that all *dharmas* (or phenomena) exist only in a subject-object interdependent framework, functioning through the five *skandhas*. There was no objective "green cup" in front of Yangsi Rinpoche during my time in class with him back at Maitripa College, discussed in Chapter 4. Instead, the subjective existence of "green cup" was a function of not just the dye molecules used to color the cup but also the observer's ability to detect and differentiate colors. As Tibetan Buddhist philosophers would put it, there is ultimately not a cup there "on its own side." The cup is indeed there, but only in a conventional sense and in the context of subject-object interdependence.

Reliance on the mythology of objectivity has consequences. It leads us to sometimes reject and ridicule nonscientific knowledge-seeking traditions in the broader Community of Knowledge Disciplines framework discussed in Chapter 6. It leads those of us working in the "hard" sciences (e.g., physics, chemistry, biology) to sometimes outright dismiss the work of "soft" sciences (e.g., psychology, sociology) where investigation of subjective phenomena is central to their scholarly missions.

Intersectionality, an innovative framework for identity, emerged from the mind of critical race theorist Kimberlé Crenshaw in the 1980s. The intersectionality concept dominates contemporary discourse in many corners of academia today, expanding from the social sciences into many diverse disciplines. Even "hard" scientists try to talk about it. Intersectionality has shaped conversations today about how to think about one's personal identity and also patterns of bias in hiring practices. In her powerful 2016 TED Talk, "The Urgency of Intersectionality,"[9] Dr. Crenshaw shared the story of her landmark legal case whereby a client was suing her employer—an automobile manufacturing company—based on a claim of discrimination, because she was a Black woman. The automaker's defense argued that the company hired women and that they also hired Black people, so the original claim was invalid. However, Dr. Crenshaw explained, all of the women employed by the company were *white* women, and all of the Black employees at the auto plant were *men*. Looking just at the individual categories—race and gender—one at a time, as many often do, was an insufficient framework that kept the real discrimination and oppression experienced by her client—a *Black woman*—invisible.

Dr. Crenshaw's genius was to create and develop the metaphor of an intersection as a lens through which her client's situation could be seen and understood. In the view of intersectionality, oppression occurs at the intersections of identity-defining categories such as race, gender, ability, sexual orientation, religion, and so on. The oppression experienced by a Black woman is unlike that of a white woman, and unlike that of a Black man. The oppression experienced by a Muslim Black woman is unlike that of a Christian Black woman, and so on. Intersectionality has become highly influential in many academic arenas and in broader society. This concept, however, has nothing to do with numbers and relies upon qualitative analysis at the intersection of categories, not the categories themselves.

A Buddhist analysis of the entrenched hierarchy of Western academia, where the "hard" sciences hold privilege over the "soft" sciences, would likely conclude that the situation had somehow ended up all wrong. Those disciplines which explicitly acknowledge and do their work in the arena of subject-object interdependence align much better with the *pratityasamutpada* Buddhist framework for reality. Thus, most Buddhists would likely assign higher epistemic authority to the "soft" over "hard" science traditions. Those operating under an ideal and/or assumption of pure objectivity would be seen as delusional by Buddhists: *samsara* stumblebummers.

Bodhi scientists critically examine themselves and their attachments to questions and hypotheses. They investigate their clinging to hypotheses, to their favorite categories, and to their numbers. In performing this form of honest critical analysis, investigators will be less blind to the impacts of their biases and emotional attachments on their science. When we see this better, we will be better able to let go of those factors and navigate wiser, more detached, and more equanimous scientific paths to truth and understanding.

Awareness

When scientists detach from illusory senses of "self" and obsessions with objects of study, their minds become more open to observations and insights that might have otherwise remained undiscovered. The history of science is laden with examples of chance discoveries and serendipity, such as the famous accidental unearthing of penicillin in 1928 by the Scottish microbiologist Sir Alexander Fleming. After returning to his lab after a two-week vacation, Sir Fleming discovered that a contaminant mold had grown on a bacterial culture plate. Upon closer examination, he noticed that the fuzzy white fungus prevented the growth of bacteria on the plate. This "happy accident" ultimately led to the dawn of antibiotics, a game-changing development that transformed biomedical approaches to combating bacterial infections.[10] A second well-known example is the serendipitous discovery of the hallucinogenic effects of lysergic acid diethylamide (LSD) by Albert Hoffman; the Swiss scientist was examining LSD for its potential to alleviate migraines and bleeding after childbirth. After ingesting a very large dose of the substance himself, he discovered through first-hand experience that the effects of the compound were something quite different.

Approximately one third to one half of scientific discoveries were made via accident or serendipity, according to a 2005 analysis by psychologists Kevin Dunbar from the University of Maryland and Jonathan Fugelsang from the University of Waterloo.[11] The ability to recognize unexpected discoveries depends upon the state of a scientist's mind. Psychological scientist Alan Baumeister from Louisiana State University described that an attentive and clever scientific mind is an essential prerequisite for a scientist to benefit from an experimental accident.[12] One of Louis Pasteur's most famous sayings was "Chance favors only the prepared mind."

The teachings of the Buddha offer ample instruction on cultivating a prepared, mindful mind. For example, the "Right Mindfulness" spoke of the Noble Eightfold Path discussed in Chapter 2 emphasizes the importance of maintaining a calm and open mind, spaciously aware of the present moment. There exist countless specific instructions and guides for mindfulness and meditation in the worlds of Buddhism, yogic, and other contemplative paths originating in Asia, as well as Western psychotherapy and wellness approaches that have borrowed extensively from the former traditions.

Buddhist meditation strategies are often subdivided into *samatha* (Pali and Sanskrit) and *vipasyana* (Sanskrit; *vipassana* in Pali). *Samatha* is frequently translated into "calm abiding" meditation and *vipasyana* is commonly considered "insight" meditation. *Samatha* quiets and stabilizes the otherwise overactive mind, making it calmer and more open and aware. *Vipasyana* meditation is analytical in nature and focuses on direct experience of core Buddhist principles such as *anitya*, *anatman*, and *pratityasamutpada*. Although *samatha* is often the first form of meditation introduced to the novice, Buddhist teachings do not prescribe *samatha* as a prerequisite to *vipasyana*. In the *Yuganaddha Sutta* of the Pali Canon,[13] the Buddha's disciple Ananda shared that those seeking enlightenment can combine *samatha* and *vipasyana* in three different ways: they can develop calm abiding and then insight, insight and then calm abiding, or they can develop the two simultaneously. Ajahn Brahm, a London-born Buddhist monk following the Theravada Thai Forest Tradition and now practicing with the Buddhist Society of Western Australia, wrote:[14]

> Some traditions speak of two types of meditation, insight meditation (*vipassana*) and calm meditation (*samatha*). In fact, the two are indivisible facets of the same process. Calm is the peaceful happiness born of meditation; insight is the clear understanding born of the same meditation. Calm leads to insight and insight leads to calm.

When one says the word "scientist," an image of a meditator does not typically come to mind. Despite the manifold benefits to different aspects of health and well-being that accompany meditation and mindfulness practice,[15] very few practicing scientists—at least in my experience—engage in such activities. The historical cultural disconnect between Buddhism and science, discussed in Chapter 4, provides one explanatory factor for this. Further, the science-mind archetype typically values narrow, laser-beam-focused analytical activity directed toward external objects of study over spacious openness and awareness, and analytical introspection. Coffee is a more frequent fuel for science minds instead of mindfulness and meditation. Most Westerners unfamiliar with meditation practice, scientist or not, sometimes initially respond to the idea of "sitting there and doing nothing" with a mix of apprehension, skepticism, and derision. This was my initial response upon being introduced to meditation at Maitripa College in 2012; I did not see the point. I just wanted to read the intriguing philosophy of Nagarjuna, Candrakirti, and others Madhyamaka thinkers without having to worry about sitting on a cushion for hours and trying not to think. After getting over my hang-ups, and going through the motions of meditation for a few months, I eventually came to deeply respect and value what it brings to all phases of my life (though it is certainly still a rollercoaster). I now better understand the essential interdependence of philosophy and practice in Buddhist paths.

A Bodhi scientist recognizes the value of an open, aware mind that is rooted in the present moment as they conduct their science. There exist many ways for scientists to cultivate balanced mental states that might be helpful to their science, such as simply going on a walk or a run, or laying down and listening to music prior to designing an experiment, collecting data, analyzing results, and so on. Centuries of Buddhist analysis suggest that mindfulness and meditation offer exceptionally promising strategies for doing good science. In my view, there is great yet vastly underexplored potential in an explicit integration of Buddhist or other contemplative practices into the scientific method. The avenues for pursuing this are numerous, with many different specific Buddhist teachings available to guide meditation and mindfulness practices and correspondingly diverse potential scientific disciplinary partners.

Before practicing science, sometimes I meditate. Before this preparatory meditation, I often turn to my favorite Buddhist advice, coming from Tilopa. He was a tenth-century Buddhist monk and philosopher from Bengal, India. Like Siddhartha Gautama, Tilopa was born into a noble life of riches which he gave up to become a wandering seeker of truth. Tilopa became a prominent figure shaping the Vajrayana ("tantric) branch of Buddhism, and his teachings became quite influential in Tibet during the centuries following his life. Tilopa is well known for a simple set of six precepts, or words of advice, that remind one to stay in the present moment, acknowledge the dynamic nature of the present moment, avoid clinging to ego ambitions and the objects of study, and to simply keep calm and openly aware. Tilopa's simple advice is:[16]

> Let go of the past
> Let go of the future
> Let go of the present
> Don't try to figure anything out
> Don't try to make anything happen
> Relax right now and rest.

Compassion

Why do we do science? Why investigate this versus that? The specific circumstances surrounding different scientists' decisions to pursue specific science paths are, of course, unique to each person, though there are also common themes. Sometimes we want to understand how things work. How do hummingbirds hover? How does DNA replicate? How do populations adapt to new environments? The aesthetic beauty of nature motivates us at other times. Our captivation with the natural world can drive us to learn as much as possible about the objects of our fascination.

One summer during my childhood, my family visited a small lake on our property in rural Missouri. It was a warm, sunny day. I sat on a large rock near the lake, mesmerized by the scene of bright blue dragonflies darting here and there around the body of water. As I was sitting there, one of the dragonflies flew right by me and hovered in front of my face for what seemed like an eternity, but in reality it was probably about five seconds. I had never experienced anything like it. It was a beautiful moment in time, and an inspiration motivating me to pursue the study of life.

For some scientists, financial and professional security is the main motivator. Some find themselves to be good at science and see this skill as a path to a good job and a good life. They might find occasional basic contentment in doing the work or merely tolerate it, but the fact that their science skills lead to professional stability, a path to paying the bills, is the main reason that they function as scientists. These three reasons, or other similar ones, are common responses likely provided by many scientists when queried about their scientific motivations.

* * *

A Buddhist perspective reveals three additional motivation categories to consider. First, though we usually don't like to admit it, scientists are often motivated by our egos, our senses of "self." We thirst to be the first, the best, and the brightest. We want the attention, admiration, and respect of others. Some, such as Jim Watson and J. Craig Venter, openly embrace and extol such self-serving scientific motivations. Some scientists, however, prefer to pretend that ego has nothing to do with it and instead assume airs of false humility. Sometimes scientists operate with an overall lack of awareness regarding how our ego motivations influence our science. We hungry ghosts of science perpetually want more and more but don't realize this to be so. These ego apparitions of science wear cloaks of false neutrality, obscuring the reality of their thirst. Such selfish motivations for pursuing science are in direct conflict with the Buddhist principle of *anatman* and are thus rooted in ignorance according to the Buddha's teachings.

Second, investigators are sometimes motivated by suffering. We are driven to seek solutions to *duhkha*. This was the case for Siddhartha Gautama: the sights of a sick person and a corpse were key experiences motivating him to embark upon the life of a renunciate and eventually achieve enlightenment. For me, in addition to my aesthetic experience with the blue dragonflies by the lake and the selfish motivations discussed in Chapter 1, human suffering also played a role in my path to the science of genetics. I had a cousin, Olivia Belle Hoover, who was beautiful. Olivia had long brown hair and a remarkable smile. Olivia, however, could not talk. She could not walk. She could not move her body at all. Sometimes Olivia smiled and seemed happy. Other times she seemed to become deeply sad, with tears streaming and looks of anguish on her face. None of us in my family knew how to help her during these times. Olivia was born with a rare genetic disorder that none of her doctors really understood. Olivia's mother, my amazing Aunt Jan, loved Olivia with the entirety of her heart. So did I and the rest of my family. Olivia died before her teen years. My Olivia *duhkha* experience was a major motivation driving me to pursue a career in genetics. I hoped to provide the world a better understanding of the nature of rare mutations that gave rise to Olivia's condition and other rare genetic diseases; I also hoped that whatever new knowledge I might provide might lead to a decrease in human suffering. Later on, after becoming a scientist and seeing my name appear on a *Science* paper, however, ego motivation took over.

Bodhicitta, the universal compassion motivation of *bodhisattvas* in Mayahana, is the third Buddhist science motivator. It is related to the *duhkha* motivation discussed earlier, but manifests in a different way. This spontaneous sense of universal and indiscriminate compassion for all beings goes beyond motivation deriving from singular experiences. *Bodhicitta* motivation also requires an understanding of Buddhist wisdom expressed through the interconnected concepts of *anitya* (impermanence), *anatman* (no-self), and *pratitysamutpada* (dependent arising, mutual cause and effect). Direct experience of these realities through meditative or other means is a prerequisite for genuine *bodhicitta*-based motivation. When we see and understand our existence in the universe as an ever-changing dynamic node, one that reflects in all directions in the infinitely interconnected network of Indra's net, we then also see and understand that anyone's suffering is our suffering. We see that the planet's suffering is our suffering, that the universe's suffering is our suffering. When we fully see and understand that we inter-are with the universe in this fashion, we become motivated to eliminate all suffering in the universe. This is the *bodhisattva* vow and the *bodhisattva* path. It is a path available to all, and it is the ultimate motivation of a Bodhi scientist.

Preventing Pseudoscience

Pseudoscience is on the rise in the COVID-19 era. The pandemic has left many people stranded in their own homes, helpless and anxious, especially during the peak "lockdown" seasons. The COVID-19 crisis, intersecting with the changing realities of an increasingly digital communication generation, created a social environment conducive to the proliferation of for new forms of pseudoscience to emerge.[17] Misinformation surrounding vaccine safety is one prominent pseudoscience thread. Echoes of eugenics and scientific racism reverberate in social media, with continuing false claims about intelligence differences between races. We see pseudoscientific claims of scientific support of fully distinct and separate binary "male" and "female" biological sex categories, which simply are not true. Dr. Mehmet Oz, an Oprah Winfrey Show regular turned pseudoscientist, cavalierly credits various dietary supplements to have biomedical benefits that are unsubstantiated, and drugs such as hydroxychloroquine to have anti-coronavirus capabilities despite a lack of credible evidence in scientific studies, for his own financial and ego gains.

Bodhi science prevents pseudoscience. Buddhist teachings, applied to scientific endeavors, provide the potential for important immunization against pseudoscientific tendencies. How can integrating Buddhist wisdom into the scientific process prevent pseudoscience? To answer this, three of

Derkson's seven sins of pseudoscience, discussed earlier in Chapter 6, will be reconsidered in light of Buddhist teachings.

The second sin of pseudoscience, *Unfounded immunizations*, connects to the Bodhi science quality of detachment. A pseudoscientist might be so attached to their ideas and hypotheses that they, knowingly or unknowingly, manipulate or alter data to make their allegedly "neutral" observations more consistent with their original claims. A Bodhi scientist, by contrast, critically examines their potential attachments to hypotheses and is better equipped to "let go" when observations are inconsistent with the original claims.

The fifth sin of pseudoscience, *Insight of the initiate*, connects to the Bodhi science quality of selflessness. Pseudoscientists often claim to bear special and unique insights into nature and reality that others do not possess. Their egos and bloated senses of "self" lead them to this conclusion. The Buddhist principle of *anatman* (no-self), by contrast, specifically identifies such views of self-importance and special insights as ignorant and misguided. A Bodhi scientist understands that they lack a "self" nature in this way and claim no special powers of insight. This is, in fact, what the Buddha taught about himself.

The sixth sin of pseudoscience, *The all-explaining theory*, also (like the second sin) connects to the Bodhi science quality of detachment. A pseudoscientist might become so attached to their ideas that they come to see their theories as universal and able to explain all. This theory attachment also leads to the pseudoscientist becoming blind to other ideas and theories that differ from their own. Thus, the sixth sin of pseudoscience also connects to the Bodhi science quality of awareness.

Pseudoscientists suffer from grandiose senses of self, severe hypothesis attachment disorders, and extreme narrowmindedness. Compassion, the fourth quality of Bodhi science, is not a consideration for pseudoscientists. The four qualities of Bodhi science presented here—selflessness, detachment, awareness, and compassion—offer guiding principles that decrease the likelihood that scientific efforts degenerate into pseudoscience, which causes profound harm to people and the societies in which they live. The Bodhi science approach also provides an investigative avenue rooted in an explicit ethical framework, the Buddha's teachings, that offers a potent and valid source of motivation for science. A Bodhi scientist is the opposite of a pseudoscientist.

10

Intimacy

Amani and the tuatara locked eyes. My son from Ethiopia, now twelve years old and garbed in Nike athletic gear from head to toe, stood still and quietly gazed into the reptile's eyes for minute after minute. Standing a few feet behind him, Stephanie and I watched this serene scene and shared a smile. Amani's hoodie hood was up, as usual; he didn't like to be noticed. I could see his thin brown nose peeking out from behind the hood and his beautiful dark eyes fixated on the tuatara. A Zealandia park tour guide told us that we were very lucky to catch a glimpse of the rare reptile; he shared that this was an adult male. The volunteer, a mustachioed twenty-something white guy named Jim originally from California, shared the sad story of how rising temperatures might make all future tuataras male and wipe out the species. I made a mental note to check into this. The tuatara remained motionless and so did Amani. Other park visitors shuffled past, unaware that the reclusive reptile was just a few feet away. I looked around to make sure Hirut was nearby; she was right behind us, completely un-interested in the tuatara. She was snapping selfies: huge eyes popping and big smile shining as she struck varying poses and gazed into her phone. A few other families walking by noticed Amani and the tuatara; the noisy younger children shuffled next to Amani, and the parents popped out their smartphones to take photos of the rare, endangered reptile. Amani and the tuatara remained quiet and still.

One mother passing by looked at Amani, and then at me and Stephanie and Hirut. A look of confusion blended with concern arose on the woman's face. She looked around a little bit, presumably seeking out Amani and Hirut's parents, and then looked back at me, the slightly bewildered look still on her face. When we made eye contact, I raised my right eyebrow and wid-ened both eyes. Her look transformed into sudden awareness and embar-rassment; the woman put her arm around her children and shuffled them along down the path.

It was time to go. Hirut snapped her last selfie, Amani said goodbye to the tuatara, and we strolled to the park café. Our transracial adoptive family continued to get more confused looks from passersby along the pathway and into the café, as is typical in our daily lives. Initial stares and quizzical looks, however, quickly dissipated into indifference or transformed into simple warm smiles. At the café, we ordered drinks and then sat at a table outside. Everyone was exhausted. When my flat white coffee arrived, I sat in simple, sipping happiness and watched the kids review the photos on their phones, pictures taken in the wildlife sanctuary during our 2019 visit to New Zealand. It had been a wonderful day.

My family brings me so much happiness. Stephanie has an uncanny ability to make everyone around her feel comfortable and appreciated. She is a veteran middle-school math teacher, now a teacher mentor, and possesses a sharp sense of logic and keen ability to help others understand things; she is the first person to whom I turn for advice. Hirut is an attention magnet and a natural leader. She notices everything and possesses a deep sense of compassion that she likes to pretend isn't there. Amani likes to fly under the radar. He is usually quiet. Sometimes, though, he seems to never stop talking. He has an infectious cackle of a laugh and an uncanny ability to hide in plain sight and then scare the living daylights out of me. The four of us have happily coevolved as a family over the last thirteen years, despite the typical *duhkha* we experience as a family with two working parents and two busy middle-school teens, and the more distinctive intersectional *duhkha* of a transracial adoptive family.

As introduced in the first chapter, for as long as I can remember, I have wanted to form a family through adoption. For the longest time, I never really understood why, but I really wanted to understand why. Across my years of life and the identity shifts along the way, from a gangly youth growing up in the Midwest to an ego-driven young evolutionary geneticist to a university professor of genetics wading in Buddhist waters, I have searched for answers as to why I wanted to adopt children instead of have biological children "of my own." Here in this final chapter, I will discuss and evaluate four explanations that I have considered along this path, and I will conclude with some final reflection on my need to undergo this introspective analysis.

Cryptic Christian Hypothesis

Many blended international adoptive families result from parents motivated by their religious beliefs. Some devout Christians are driven to adopt children from a religion-based moral basis, rooted in their faith in a creator God. Christians often cite Bible verses as their justifications or rationales for particular beliefs or actions, adoption no exception:

> Religion that is pure and undefiled before God the Father is this: to visit orphans and widows in their affliction, and to keep oneself unstained from the world. (James 1:27, English Standard Version)

In the early 2000s, American evangelicals became a powerful force in international adoption, commanding their devout believers to follow the biblical mandate that Christians care for "orphans and widows in their distress."[1] In 2007, celebrity pastor and megachurch mogul Rick Warren, along with other evangelical Christian leaders, encouraged their followers to turn their focus toward adoption. This call was, in part, a response to pro-choice voices that challenged pro-life Christians to "adopt all the babies they wanted to be born." In 2009, the Southern Baptist Convention passed a resolution directing all members to reflect on whether God was calling on them to adopt. Christian families began adopting from international nations where poverty was prevalent and the children were Black, such as Ethiopia and Haiti. The movement began to refer to adoption as a means of "redeeming orphans"; their original families became either completely forgotten or cast in negative lights by the Christian (and usually white) adopters. Religious adoption agencies and ministries often portrayed the birth mothers as hopeless, promiscuous addicts, and bad influences from whom the children's souls needed saving.

Scandals soon emerged. In some cases, the presumed orphans being placed for adoption were not in fact orphans at all. One well-known case involved a Tennessee couple who adopted a young girl from Ethiopia, but when the adoptee had learned enough English to communicate to her new family, she told the new parents that she in fact already had a mother back in Ethiopia.[2] The Christian couple was under the impression that the girl had no living parents. The couple called the agency to demand an answer, and the child's claim was confirmed. The adoptive parents were devastated, feeling as if they had stolen someone else's daughter, and embarked upon an arduous journey to find the girl's biological mother, which was ultimately unsuccessful.

In the early 2010s, some adoptive families got so caught up in the evangelical Christian mission that they adopted as many children as possible, with disastrous consequences sometimes accompanying these adoptive mega-families. Some adoptees were abused or even killed at the hands of their adoptive parents.[3] The headlines surrounding these stories resulted in a massive regulatory curtailing of adoption from Ethiopia and a drastic overall reduction in international adoption. Despite these highly publicized horror stories, most international adoptions involving Christian parents have presumably not been marked by such tragedy, but rather followed more typical paths of adoptive families dealing with typical joys and struggles, day-to-day *duhkha*.

Given the recent prominence of international adoption among some Christian communities, one hypothesis to explain my motivation to form a family through adoption might involve Christianity-influenced beliefs and values. I grew up in a small town in Missouri where Christianity was (and remains) prominent. However, I have never truly identified as a Christian at any point in my life. As a youth, my family would celebrate Christmas and on very rare occasion attend a local church on Easter, but otherwise Christianity was never really a significant part of my upbringing. Opportunities to join the faith presented themselves at Young Life gatherings and church-organized sports camps that I attended with friends during my teen years. However, this religious tradition's teachings and scriptures never found resonance with my sense of reason. For example, at one Young Life camp at the age of fifteen, I interrupted the youth leader during one of his discussions to ask how Christianity dealt with the fact that the Bible must have been translated many times to arrive in the particular English version of the text which he held in his hand. The leader responded that God was watching over the translations. Others around me nodded their heads and seemed satisfied with this answer, but I was not. My dissatisfaction with Christianity and its explanations for how things are and why certain things should be done (or not done) cemented during my college years, when I was introduced to Darwinian evolution in an introductory biology class, along with the stories of its historical clashes with Christianity. I switched my major from chemistry to biology, and began reading books by Richard Dawkins and Stephen Jay Gould. I was inspired by these evolutionary thinkers. Thus, the virtually complete absence of meaningful exposure to Christianity, advancing to derisive attitudes toward the religion during my teens and twenties, presents sufficient evidence to discard a Christian "soul saver" or other similar hypothesis as a prominent factor contributing to my desire to form a family through adoption.

Evolutionary Hypotheses

In 2010, three years after adopting Amani and Hirut, I served on a National Science Foundation (NSF) grant proposal review panel in Washington, DC. The panel focused on applications coming from researchers in the field of evolutionary genetics and genomics. As a junior professor, this was my first time serving on such a board, and I was honored to be included in this select group charged with evaluating many dozens of outstanding proposals from scientists around the nation. The panel work was very high stakes; we all knew that only ~10 percent of the many outstanding research proposals would ultimately be selected for funding. The experience was exhausting. The group met across three days, from ~8:00 am to 7:00 pm each day, carefully evaluating each proposal under consideration.

After the third grueling day, the NSF panel director rewarded the review group with dinner and drinks. We met at a nearby Italian restaurant, and after settling into chairs, everyone enthusiastically ordered the drink of their choice when the server arrived. A basketball game was on the TV in the corner by the bar; I happily zoned out for a little bit watching the Washington Wizards get blown out by the Boston Celtics. My beer glass quickly became empty, and my mind was at rest. The Wizards took a much-needed timeout, and then a commercial for an upcoming entertainment news show appeared on the screen, showing Brad Pitt and Angelina Jolie on vacation with their large multiethnic family, which included adoptees from Cambodia, Ethiopia, and Vietnam. I then heard a loud groan next to me. The grumble came from a fellow panelist occupying the seat to my left, a fellow bespectacled and white junior professor; I hadn't interacted with him much. He noticed me watching the TV and then nudged his chair closer to mine. He leaned in, pointed to the famous family on the screen, and looked at me; with a wry smile and condescending tone, he said, "Isn't that just so pathetic?" The young professor proceeded to ridicule the popularized international adoption trend, noting how all of those adoptive parents were so ignorant and misguided. He wondered why Brad and Angelina would waste their time on raising those defective children when offspring of their own would certainly be remarkable.

I took two deep breaths and excused myself to go to the bathroom. I sat on a toilet seat for a few minutes, trying to deal with my rage. I eventually calmed down. When I came back, I happily saw that the asshole had turned his attention to other panelists; I took my seat, ordered another beer, and remained quiet for the rest of the dinner. In my hotel room later that night,

I cried. I felt weak and impotent, so ashamed that I retreated to the bathroom instead of confronting that jerk. I reflected on the fact that I didn't even know how to defend the beautiful and altruistic act of adoption, and my beautiful family.

* * *

Altruism is often framed as a "paradox" in evolutionary biology. The occurrence of biological altruism, behaviors and actions that increase the reproductive fitness of another individual with no benefit to the actor, has elicited a range of evolutionary explanations for this widely observed phenomenon. Biological altruism is not be confused with psychological (or philosophical) altruism; the latter involves caring about others' welfare and deliberately benefiting them for their own sake, with no restriction on the type of benefit involved. By contrast, biological altruism has nothing to do with intentions or motives, and it pertains only to the biological "benefits" to others that increase their reproductive fitness, thus increasing their genetic contribution to future generations.

One well-known theory for explaining biological altruism is kin selection. The general idea of this theory is that altruistic behaviors are expected when genetic relatedness is high. In a mathematical model of kin selection, known as "Hamilton's Rule," natural selection causes alleles (versions of genes) to increase in population-level frequency when the genetic relatedness of a recipient to an actor, multiplied by the benefit to the recipient, is greater than the reproductive cost to the actor.[4] Group selection provides a second and more controversial evolutionary theory; here, natural selection acts at the level of the group, instead of at the more conventional and accepted level of the individual organism. During the second half of the twentieth century, influential evolutionary biologists such as John Maynard Smith and Richard Dawkins criticized group selection, deploying various mathematical approaches to argue that individuals would not altruistically sacrifice their own fitness for the sake of a group.[5,6] In 1994, David Sloan Wilson and Elliott Sober argued in favor of group selection on the basis of the ability of selection to work on many levels, not just that of the individual gene or organism.[7] A controversial 2010 paper, including the eminent E. O. Wilson as an author, argued that group selection can occur when competition between two or more groups, some containing altruistic individuals who act cooperatively together, is more important for survival than competition between individuals within each group.[8] This paper provoked a strong rebuttal from

103 evolutionary biologists.[9] Though increasing in acceptance, group selection-based explanations for altruistic behaviors in natural populations remain controversial and divisive in evolutionary circles.

Some have argued that the evolution of kin selection–based biological altruism gave rise to psychological altruism. In his article, "Morality and Evolutionary Biology," William FitzPatrick explained this point of view with the following example:[10]

> . . . suppose you receive a letter from UNICEF soliciting contributions for health and nutrition programs for children in Darfur, and you are moved to send a check. This is not merely selective altruism toward kin or likely reciprocators, but altruism toward strangers who are in no position to reciprocate, and it might therefore seem puzzling from a purely biological point of view: such "indiscriminate altruism" isn't biologically adaptive in the way more selective altruism might be; your helping children in Darfur isn't helping to propagate your own genes, so it may seem mysterious how such a trait could have evolved through natural selection. But a trait that is not presently adaptive may once have been. In the environment in which our hominin ancestors lived, where there was little positive contact with outsiders, even relatively indiscriminate altruism would tend to benefit kin or potential reciprocators, and so might have been a simple adaptive mechanism on the whole.

"Indiscriminate altruism," according to this line of thinking, reflects historically relevant adaptations that are for the most part no longer relevant. This theory frames such indiscriminate altruistic acts, such as donations to UNICEF or adoption of children, as a sort of "evolutionary misfiring" of formerly adaptive behavioral instincts, as portrayed by Richard Dawkins in *The God Delusion*.[11] In his article,[12] FitzPatrick noted the speculative and limited nature of such evolutionary explanations, considering that indiscriminate altruistic tendencies have nothing really to do with evolutionary adaptation but are instead founded upon the values that an individual person has developed through social and cultural contexts and personal reflection.

My identity as a Richard Dawkins superfan started to decline in 2007, the year I read *The God Delusion* and traveled to Ethiopia with Stephanie to adopt Amani and Hirut. Upon buying the book, I was so excited to read it, anticipating the pointed and witty arguments against religion for which Dawkins was renowned. By this time, however, I had also started reading books about

Buddhism, just a few years into my ongoing journey investigating this way of thinking. I finished Dawkins's then-new book right before we left for Ethiopia. It was long, more than four hundred pages, and although some of the arguments were sharp and well founded, other parts of the book seemed to unjustifiably overextend the reach of natural selection. The "evolutionary misfiring" explanation of indiscriminate altruistic traits offered one example; another was Dawkins's extension of biological natural selection principles to the formation of stars and other astronomical processes. My long-standing personal desire to adopt was certainly an element contributing to my dissatisfaction with the "misfiring" hypothesis, particularly as my trip to adopt from Africa was just around the corner. Further, to me, the overly long reach of Darwinian evolution presented in Dawkins's latest book echoed "the all explaining theory," Derkson's sixth sin of pseudoscience.

From a more scientific perspective, I was also troubled by the cavalier extensions of purely natural selection–based arguments to phenomena such as charitable giving and galaxy formation, without first considering nonselective explanations. Just two years earlier, I was a postdoctoral trainee under the mentorship of Michael Lynch, who was (and remains) a master of identifying, unraveling, and often debunking selection-based arguments for various biological phenomena that failed to first consider nonselective evolutionary explanations. In *The Origins of Genome Architecture*, for example, Lynch offered a thorough and compelling argument that the evolution of large and complex genomes in multicellular organisms, such as humans and trees, was primarily a result of nonadaptive processes such as genetic drift and mutation.[13] Lynch was composing this book and related work during my years under his mentorship at Indiana University; as a consequence of this training and exposure, I became conditioned to immediately respond to any and all adaptive explanations for whatever evolutionary phenomenon with reservation and doubt. Despite my years of immersion in the theories and practices of evolution and genetics, evolutionary explanations for my motivation to pursue a family through adoption have remained, for me, inadequate.

White Savior Hypothesis

Is it all white guilt? Amani and Hirut are, after all, from Ethiopia. They are Black. I am white. Does my long-standing desire to adopt children, and the ultimate decision to do so specifically from Ethiopia, reflect a deeply seeded psychological guilt over my white roots and the centuries of atrocities and oppression inflicted upon Black people from Africa, by people of European ancestry who looked like me?

I often reflect upon a memory of an eleven-year-old version of me, in 1985, sitting in front of the family room TV and watching "We Are the World." This famous music video featured many iconic musicians of the 1980s who sang a series of short successive solos. The song featured Lionel Richie, Stevie Wonder, Paul Simon, Kenny Rogers, James Ingram, Tina Turner, Billy Joel, Michael Jackson, Diana Ross, Dionne Warwick, Willie Nelson, Al Jurreau, Bruce Springsteen, Kenny Loggins, Steve Perry, Daryl Hall, Huey Lewis, Cyndi Lauper, Kim Carnes, Bob Dylan, and Ray Charles. Many millions of dollars in donations and record sales resulted from this phenomenon, and the funds were donated to the nonprofit United Support of Artists (USA) for Africa. The money went to purchase grain to help feed and relieve starving people in Africa, especially those in Ethiopia, where approximately one million people died during the country's 1983–1985 famine. This TV-watching experience impacted the version of me that existed back then, a skinny white preteen who knew virtually nothing else of Africa or any other part of the world outside the northwest corner of Missouri in which I grew up. "We Are the World" shaped my early and naïve views of Africa and its people, like countless others in America at the time, painting a picture of a vast continent of countless starving Black people who desperately needed our white help.

* * *

A quarter century after "We Are the World," another video directed American attention to Africa. *Kony 2012* was a short documentary, produced by Invisible Children, Inc. of San Diego. The purpose of this film was to expose the acts of Joseph Kony, a militia leader from Uganda, and avenge his victims. The film described Kony's actions with his rebel militia group, the Lord's Resistance Army (LRA), which forcibly recruited child soldiers and was accused of numerous war atrocities. One of the central figures featured in the film is a young Ugandan boy named Jacob Avaye, whose brother was killed by the LRA. Jason Russell, a white film artist from southern California who directed and appeared in the documentary, promised Jacob that he would help "stop Kony." *Kony 2012* was the first video ever to reach one million likes on YouTube, and it was named the most viral video ever by TIME Magazine that year.[14] Some hailed Invisible Children's social media approach as a transformative event in global justice. For example, Anneke Van Woudenberg from Human Rights Watch wrote:[15]

> We've spent years investigating the horrors perpetrated by the LRA in central Africa—Uganda, Democratic Republic of Congo, Central African Republic (CAR), and South Sudan. We gathered evidence at massacre sites—wooden clubs covered in dried blood, rubber strips from bicycle tires used to tie up the victims, and freshly dug graves—and spoke to hundreds of boys and girls forced to fight for his army or held captive as sex slaves. And we're elated that #stopKony is a trending topic on Twitter—if anyone deserves global notoriety it's Kony.

Others, however, criticized Jason Russell and Invisible Children, Inc.'s approach. On March 8, 2012, Teju Cole, an African American novelist and social commentator who holds the Gore Vidal Professor of the Practice of Creative Writing chair at Harvard University, posted a series of seven tweets after he watched *Kony 2012*:

1. From Sachs to Kristof to Invisible Children to TED, the fastest growth industry in the US is the White Savior Industrial Complex.
2. The white savior supports brutal policies in the morning, founds charities in the afternoon, and receives awards in the evening.
3. The banality of evil transmutes into the banality of sentimentality. The world is nothing but a problem to be solved by enthusiasm.
4. This world exists simply to satisfy the needs—including, importantly, the sentimental needs—of white people and Oprah.
5. The White Savior Industrial Complex is not about justice. It is about having a big emotional experience that validates privilege.
6. Feverish worry over that awful African warlord. But close to 1.5 million Iraqis died from an American war of choice. Worry about that.
7. I deeply respect American sentimentality, the way one respects a wounded hippo. You must keep an eye on it, for you know it is deadly.

These tweets, like the Kony 2012 video that sparked them, also went viral. Teju Cole received support from some, but others called him a racist and a Mau Mau. In the first tweet, Cole dissed Nicholas Kristof, a renowned (and white) journalist who won two Pulitzer Prizes, widely known for shedding light on global social injustices such as human trafficking and the Darfur tragedy. Archbishop Desmond Tutu of South Africa hailed Kristof as an "honorary African" for shining a spotlight on overlooked African conflicts.[16] In an *Atlantic* article published just fourteen days after his to-become-viral tweets were posted, Teju Cole wrote:[17]

> I want to tread carefully here: I do not accuse Kristof of racism nor do I believe he is in any way racist. I have no doubt that he has a good heart. Listening to him on the radio, I began to think we could iron the whole thing out over a couple of beers. But that, precisely, is what worries me. That is what made me compare American sentimentality to a "wounded hippo." His good heart does not always allow him to think constellationally. He does not connect the dots or see the patterns of power behind the isolated "disasters." All he sees are hungry mouths, and he, in his own advocacy-by-journalism way, is putting food in those mouths as fast as he can. All he sees is need, and he sees no need to reason out the need for the need.

Cole went further to explain that the actions of those such as Kristof and Invisible Children, Inc., though well intentioned, neglected to afford agency to Africans living in places such as Uganda. He pointed out that there are many African journalists and writers who have done important work in bringing to light the LRA and other sources of injustice in Africa, and that African writers usually see the situations with a clearer and more direct vantage point. By contrast, "white saviors" often present the circumstances in incomplete and/or inaccurate ways. He alluded to statements such as "We have to save them because they can't save themselves" and other sentimental though misguided comments churned out from the cogs of the White Savior Industrial Complex.

So was my desire to adopt children from Ethiopia a mere reflection of white guilt, just another example of misguided action resulting from the White Savior Industrial Complex? Does Teju Cole's fifth tweet explain things . . . that I was aiming to have a big emotional experience to validate my white privilege? I initially met this and similar "white guilt" explanations with obstinance and feelings of personal offense, typical of white defensiveness when matters of race arise. After extensive reflection I reached a conclusion, speaking just for myself, that white guilt was indeed a contributing factor that influenced my personal desire to form an adoptive family. My 1980s youth experience sitting in front of a TV and repeatedly watching the "We Are the World" music video, and the surrounding programming that showed pictures of starving children in Africa, was an undeniable influence. Sally Struthers's 1980s TV commercials for the Christian Children's Fund showed similar imagery of malnourished African children and pleas for financial help. Experiences such as this, just like all of my life's experiences, contributed to the mutual cause-and-effect underpinning that shaped who I am and the decisions I make.

I have come to be OK with this reality, and despite it, I remain confident and happy in my family's transracial adoption path and how it has changed my universe. My family's experiences have motivated me to better understand systems of oppression and structural racism throughout America and its institutions, and to do my best to change these systems. In alternate universes where I was not an adoptive father to African children, it is highly unlikely that I would have become aware of and influenced by the works of Teju Cole, Kimberlé Crenshaw, Ta Nehisi Coates, and other Black intellectuals of our time. Raising Amani and Hirut has made me acutely aware that I am very much a white man, through and through, and that I will never directly experience or truly understand what my children go through.

White guilt provides a valid variable contributing to my decision to pursue adoption from Africa, but this factor alone is insufficient to explain my foundational desire to form a family through adoption in the first place. "Saving the poor children from Africa" was never a motivator or a component of discussions that Stephanie and I had in our adoption conversations. When we started thinking seriously about adoption after moving to Oregon in 2006, we considered many international possibilities, including China, Russia, Honduras, Ghana, and Ethiopia. One key circumstantial factor that drove us to decide on Ethiopia was the fact that there was an adoption agency not far away in Banks, Oregon, which had a long-standing relationship with an orphanage in Addis Ababa. Stephanie and I were very impressed with the people working at the agency, and we heard good things from other adoptive families in our community that worked with them.

Rather than white savior complex, I was more guilty of an erroneous colorblind mindset (I remember thinking at the time: "It doesn't matter where our children come from . . .") typically held by many white people, including liberal academics such as me, before the so-called racial reckoning that followed the murders of Ahmaud Arbery, George Floyd, Breonna Taylor, and other 2020 Black victims of police and other white violence. The version of me pursuing adoption back then was profoundly naïve about the impact that my children's racial identity would have on them, and even more naïve about the profound relevance of my different, privileged racial identity in this familial context. I now know and understand through direct experience that the color of my children's skin, and my skin, matters a lot in many different ways. I acknowledge the effects of white savior complex and an erroneous color-blindness mindset in shaping our specific adoptive path. However, such explanations alone are insufficient to fully explain my long-standing, deep-down desire to form a family through adoption instead of through biological reproduction, especially in those years long before the racial identity of our children was even a matter to consider.

Buddhist Understanding

My privileged access to the teachings of the Buddha and other Buddhist thinkers revealed three key insights into my desire to form a family through adoption, and my desire to understand this. The first of these derives from *pratityasamutpada*, the Buddha's mutual cause-and-effect framework. This principle describes how the universe works, and the concept has been further developed and elaborated upon by Buddhist thinkers across the last twenty-five centuries. *Pratityasamutpada* extends to me and the transracial adoptive family where I am a father. A *pratitysamutpada*-based analysis reveals that, like all other phenomena, there were multiple causes leading to my desire to adopt and the circumstances of how it all played out and continues to play out. My mother and father, Kay and Phil Denver, instilled in me a basic value of kindness and helping others throughout my childhood years. Even though I have never held a Christian faith, watching Sally Struthers's commercials for the Christian Children's Fund during my youth affected me. Awareness of the "evolutionary misfiring" hypothesis shaped my thinking and attitudes toward adoption. White guilt and naïve color-blindness played a role. There are countless other causes and conditions as well that led to the realization of my adoptive family's continuing path.

The second Buddhist insight that contributed to my understanding of the adoption question is found in *bodhicitta*, the spontaneous sense of universal compassion for everyone, everywhere. *Bodhicitta* is the Buddha's rationale for indiscriminate kindness. This concept, central to Mahayana Buddhist paths, is the underlying motivation for *bodhisattvas* and their vows to forego *nirvana* until all sentient beings in the universe are enlightened. *Bodhicitta* cannot be quantified, and it is said to be knowable only through direct, subjective experience. Throughout my years, I have experienced brief glimmers of *bodhicitta*—sometimes while sitting in meditation, though more often it comes out of nowhere during day-to-day experiences. *Bodhicitta* was there when I locked eyes with the Buddhist monk in Bloomington, and when I serenely watched Amani and the tuatara lock eyes in New Zealand. When I first learned of this concept, sitting in the Maitripa College classroom in 2012, my mind immediately connected dots between *bodhicitta* and my desire to adopt.

Bodhicitta underpins the Buddha's compassion-based ethical framework. Buddhism, like the other major spiritual traditions of the world, is a proponent of the famed Golden Rule: do unto others as you would like them to do unto you. A selfish and materialism-minded cynic, however, might respond, "Why should I? What's in it for me?" The Buddha's teachings offer a uniquely powerful response, rooted in the logic of *anatman* and *pratityasamutpada*: the reason why you should do unto "others" as you would like them to do unto "you" is because in reality nothing separates the "others" from "you." Any action you take on the "other" you will also necessarily take upon your "self." Any action you take on your "self" you also take on the "other." As the Vietnamese monk Thich Nhat Hanh would put it, the "self" and the "other" inter-are.

When I breathe in oxygen molecules from the air, it is important to remember that those were once part of a tree. When I exhale carbon dioxide out into the air, those molecules will become a tree in the future. In this way, and countless other ways, we are all interconnected in molecular intimacy, inescapably interwoven into Indra's net. Under this line of reasoning, indiscriminate compassion is the only logical option. Knowing *bodhicitta* helped me understand that my love and compassion are equally available to anyone and everyone, regardless of the inheritance paths of DNA.

Third, Buddhist teachings reveal the totally unnecessary nature of this entire line of questioning and hypothesizing about the reasons why I wanted to adopt. Candrakirti, the seventh-century Madhyamaka philosopher discussed in Chapter 9, would say that my busy mind's need to articulate and investigate the question reflects my limited understanding of the *dharma* in the first place. If I asked Siddhartha Gautama what he thought the source of my desire to adopt might be, the Buddha would likely have no response. My family brings me love and *bodhicitta*. That is all I really need to know.

* * *

The rain became the tree, and the tree became the rain. The rain became the Buddha, and the Buddha became the rain. The rain became DNA, and DNA became the rain. The rain became Amani, and Hirut, and Stephanie, and me.

Notes

Chapter 1

1. The Buddha's teachings became known to the ancient Greeks through the emissaries of Asoka, the Buddhist emperor of the Maurya dynasty who ruled most of the Indian subcontinent from ~268 to 232 BCE. Buddhism also flourished under the Indo-Greek and Greco-Bactrian kings from ~200 to 135 BCE. In modern times, the Buddha's teachings have become more widely known and popularized throughout Europe and the Americas through the recent activities of Buddhist teachers such as D. T. Suzuki, the Dalai Lama, Ayya Khema, and Thich Nhat Hanh.
2. D. R. Denver, K. Morris, M. Lynch, L. L. Vassilieva, and W. K. Thomas, "High Direct Estimate of the Mutation Rate in the Mitochondrial Genome of *Caenorhabditis elegans*," *Science* 289 (2000): 2342–2344.
3. D. R. Denver, K. Morris, M. Lynch, and W. K. Thomas, "High Mutation Rate and Predominance of Insertions in the *Caenorhabditis elegans* Nuclear Genome," *Nature* 430 (2004): 679.
4. J. Palca, "Evolution, Mutation May Occur Faster Than Thought," *All Things Considered*, National Public Radio, August 4, 2004.
5. M. Lynch, "Mutational Meltdowns in Sexual Populations," *Evolution* 49 (1995): 1067–1080.
6. M. Lynch and J. S. Conery, "The Origins of Genome Complexity," *Science* 302 (2003): 1401–1404.
7. R. Dawkins, *The Selfish Gene* (New York: Oxford University Press, 1976).
8. R. Dawkins, *The Extended Phenotype: The Long Reach of the Gene* (New York: Oxford University Press, 1999).
9. His Holiness the Dalai Lama, *How to Practice: The Way to a Meaningful Life*, ed. and trans. J. Hopkins (New York: Atria, 2002).
10. In referring to the socially constructed race categories in America, I will follow the *New York Times* style guide, updated shortly after the murder of George Floyd. "Black" will be capitalized throughout the text to recognize and emphasize the shared cultural identities of those individuals who experience being Black in America. By contrast, the "w" in "white" will remain lowercase due to the widespread usage of capitalized "White" among white supremacists in their racist literature and propaganda. The style update and rationale are detailed in N. Coleman, "Why We're Capitalizing Black," *New York Times*, July 5, 2020.

Chapter 2

1. A 2009 PBS documentary by David Grubin, *The Buddha*, offers a particularly compelling and complete, yet succinct account of the Buddha's life.

2. Bhikku Namamoli and Bhikku Bodhi, trans., *The Middle Length Discourses of the Buddha, a Translation of the Majjhima Nikaya*, 4th ed. (Somerville, MA: Wisdom Publications, 1995).

3. *Bodhisatta* (Pali) is a term used to describe the Buddha before his enlightenment in the *suttas* of the Pali Canon. The Sanskrit version of this term (*bodhisattva*), prominent in Zen and Tibetan forms of Buddhism, refers more broadly to those following prescribed paths to enlightenment.

4. In Theravada, *sangha* generally refers to the ordained monastic community; in Zen and other Buddhist traditions, *sangha* sometimes more broadly indicates any community of practitioners.

5. Brahmin here refers to a Vedic priest cast in ancient Indian society.

6. Though Sariputta is portrayed as wise and enlightened in the Pali Canon of Theravada, in the *sutras* of Tibetan and Zen traditions, this individual (Sariputra in Sanskrit) is often depicted in unflattering fashions, frequently asking questions that reveal an incomplete or unsophisticated understanding of the *dharma*.

7. The name of the wanderer to which the Buddha spoke was Dighanakha; "Aggivessana" refers to the clan of people from which this wanderer came, the Agnivesyayanas.

8. Walpola Rahula (1907–1997) was a Sri Lankan Theravada Buddhist monk and scholar who also studied other forms of Buddhism at Calcutta University and the Sorbonne in Paris. He became the Professor of History and Religions at Northwestern University in 1964, thus becoming the first ordained Buddhist monk to hold a professorial chair in the Western world.

9. W. Rahula, *What the Buddha Taught* (New York: Grove Press, 1959).

10. MN 141.22.

11. MN 117.3.

12. Bhikku Bodhi, trans., *Numerical Discourses of the Buddha: A Translation of the Anguttara Nikaya* (Somerville, MA: Wisdom Publications, 2012).

13. *Bhikku* (Pali; *bhiksu* in Sanskrit) is a term to describe a male Buddhist monk. A female Buddhist monk is called a *bhikkuni* (Pali; *bhiksuni* in Sanskrit).

14. MN 44.11.

15. The Buddha further provided a detailed exposition of *pratityasamutpada* in this *sutra*, involving a linear causal chain composed of twelve factors, connected together by eleven propositions, which explain recurring rebirth in the *duhkha* of *samsara*. The Buddha continued in the *Bahudhatuka Sutta*: "That is, with ignorance as condition, formations [come to be]; with formations as condition, consciousness; with consciousness as condition, mentality-materiality; with mentality-materiality as condition, the six-fold base; with the six-fold base as condition, contact; with contact as condition, feeling; with feeling as condition, craving; with craving as condition, clinging; with clinging as condition, being; with being as condition, birth; with birth as condition, aging and death, sorrow, lamentation, pain grief, and despair come to be. Such is the origin of this whole mass of suffering."

16. Piyadassi Thera, *Dependent Origination* (Paticca Samuppada), The Wheel Publication No. 15 (Kandy, Sri Lanka: Buddhist Publication Society, 1959).

17. N. Samten and J. L. Garfield, trans., *Ocean of Reasoning: A Great Commentary on Nagarjuna's* Mulamadhyamakakarika (New York: Oxford University Press, 2006).

18. S. Batchelor, *Buddhism Without Beliefs: A Contemporary Guide to Awakening* (New York: Riverhead Books, 1997).

19. S. Batchelor, *The Awakening of the West: The Encounter of Buddhism with Western Culture* (Berkeley, CA; Parallax Press, 1994).

Chapter 3

1. M. Walshe, trans., *The Long Discourses of the Buddha: A Translation of the Digha Nikaya*, 2nd ed. (Somerville, MA: Wisdom Publications, 1995).

2. The specific number of early schools is debated by scholars; some accounts suggest the presence of twenty instead of eighteen such schools.

3. Some scholars consider Vajrayana to be a distinct third main branch of Buddhism; others consider it to be a sub-branch of Mahayana. Vajrayana—often referred to as "tantric" or "esoteric" Buddhism—was a prominent influence in Tibet, Bhutan, and Mongolia. The main differences between Vajrayana and other forms of Mahayana are methodological in nature, with the tantric methods of Vajrayana offering a faster vehicle to enlightenment as compared to the slower methods of other Mahayana branches.

4. Though the Sthavira School, forerunner to modern Theravada Buddhism, was among those eighteen early schools that Mahayana Buddhists considered as "Hinayana."

5. The *Ksitigharba Sutra* related a fantastical story of a young girl, Ksitigharba, whose mother died and was destined for rebirth in a hell realm because the woman frequently slandered the Buddha's teachings. Ksitigharba sold all of her possessions and prayed at a local temple, where she received guidance from the Buddha on how to save her mother from rebirth in the hell realm. She followed the instructions and discovered that her mother had in fact been saved from the torturous rebirth that she feared. However, upon passing through the hell realm, Ksitigharba observed countless individuals in profound agony. This experience cultivated *bodhicitta* in the young girl, and she subsequently took the vow to become a *bodhisattva* and liberate all beings from *samsara*.

6. Later forms of Mahayana added four more steps to bring the list to ten, though the six original *paramitas* is most common. Theravada Buddhism has its own set of ten perfections, with many points of overlap with the Mahayana list.

7. E. Conze, *The Perfection of Wisdom in Eight Thousand Lines & Its Verse Summary* (San Francisco, CA: The Four Seasons Foundation, 1973).

8. Though *sunyata* (Sanskrit), translated as "emptiness," is most commonly associated with Mahayana Buddhism, *sunnata* (Pali) is also prevalent in Theravada Buddhism but is instead translated as "voidness" and refers to the void aspect of *nibbana* that an *arahant* detects and recognizes through meditative analysis. In the Pali Canon, the *Culasunnata Sutta* (MN 121) and *Mahasunnata Sutta* (MN 122) offer advice on seeking seclusion necessary to realize *sunnata*.

9. The translation used here derives from the version used at Maitripa College, translated from the Tibetan.

10. J. L. Garfield, *The Fundamental Wisdom of the Middle Way: Nagarjuna's Mulamadhyamakakarika* (New York: Oxford University Press, 1995).

11. Tsongkhapa and followers of the Prasangika-Madhyamaka school favored a "consequential" debate strategy whereby monks would respond to the claims or theses of their opponents through *reductio ad absurdum* and similar approaches, but never set forth their own theses. This contrasted with the Svatantrika-Madhyamaka, or "syllogistic," school that deployed autonomous syllogistic approaches (i.e., using stated premises) in debate. Prasangika-Madhyamaka adherents claimed that the use of such approaches reflected a subtle form of believing in inherent features of the targets of debate analysis.

12. Translated from the Tibetan by Geshe Thupten Jinpa.

13. J. L. Garfield, "Taking Conventional Truth Seriously: Authority Regarding Deceptive Reality," in *Moonshadows*, Ed. The Cowherds (New York: Oxford University Press, 2011), 29–30.

Chapter 4

1. C. I. Beckwith, *Greek Buddha: Pyrrho's Encounter with Early Buddhism in Central Asia* (Princeton, NJ: Princeton University Press, 2015).

2. T. McEvilley, *The Shape of Ancient Thought: Comparative Studies in Greek and Indian Philosophies* (New York: Allworth Press, 2012).

3. R. Bett, *Pyrrho, His Antecedents, and His Legacy* (New York: Oxford University Press, 2000).

4. B. A. Wallace, ed., *Buddhism & Science: Breaking New Ground* (New York: Columbia University Press, 2003).

5. His Holiness the Dalai Lama, *The Universe in a Single Atom: The Convergence of Science and Spirituality* (New York: Broadway Books, 2005).

6. J. K. P. Ariyaratne, *Two Buddhist Sutras Viewed from Science* (Pannipitiya, Sri Lanka: Stamford Lake, 2003).

7. This tree, "Sri Maha Bodhi," is said to still live on today in Anuradhapura, and it is considered the oldest historical tree on the planet with a planting date of 288 BCE. The tree is commonly said to be the "southern branch" of the original Bodhi tree under which the Buddha sat during his enlightenment.

8. In 1017 CE, the Aunradhapura Kingdom of Sri Lanka fell to the Chola Empire of southern India, and both the *bhikku* and *bhikkhuni* orders in Sri Lanka disappeared. King Vijayabahu of Sri Lanka drove away the Cholan invaders in ~1070 and revived the Theravada *bhikku* order with monks from Burma. The *bhikkuni* order, however, remained unrestored in Sri Lanka.

9. For example, in 1864, it was decreed that every child born in Sri Lanka must be given a Christian name.

10. C. Darwin, *On the Origin of Species by Means of Natural Selection, or the Preservation of Favoured Races in the Struggle for Life* (London: John Murray, 1859).

11. This quotation was found in D. S. Lopez Jr., *Buddhism & Science: A Guide for the Perplexed* (Chicago, IL; University of Chicago Press, 2008).

12. Lopez, *Buddhism & Science*.

13. T. H. Huxley, "II. Agnosticism," in *Christianity and Agnosticism: A Controversy* (New York: The Humboldt Publishing Company, 1889), 97.

14. T. H. Huxley, *Evolution and Ethics and Other Essays* (London: Macmillan and Company, 1894).

15. The Sinhalese are an ethnic group, originally from northern India, that today make up approximately 75 percent of the population of Sri Lanka and are predominantly Theravada Buddhist.

16. Patricia Lee Masters, an ordained Buddhist nun and scholar of diverse Buddhist traditions who passed in 2018, received a Fulbright award to perform research on Anagarika Dharmapala and Mary Foster, a Hawaiian associate of Dharmapala and champion of Buddhist communities on the islands of Hawaii. Masters studied Dharmapala's diaries and other resources in 1999 during her Fulbright sabbatical. She wrote of these experiences and resources in P. L. Masters, *Searching for Mary Foster* (New York: American Buddhist Study Center, 2017).

17. Quotation found in Lopez, *Buddhism & Science*. Lopez drew on *A Full Account of the Buddhist Controversy, Held at Pantura, in August, 1873. By the "Ceylon Times" Special Reporter: with the Addresses Revised and Amplified by the Speakers* (Colombo, Sri Lanka: Ceylon Times Office, 1873).

18. G. J. Melton, ed., "Theosophical Society," *New Age Encyclopedia* (Farmington Hills, MI: Gale Research, 1990).

19. In *Buddhism & Science*, Lopez notes that the name chosen by Don David was not Pali in origin (the Theravada Buddhist language of his native Sri Lanka), but rather Sanskrit. This might reflect the influence of Madame Blavatsky, who frequently used Sanskrit terms in her writings.

20. Cited in Lopez, *Buddhism & Science* and Masters, *Searching for Mary Foster*.

21. Many of Dharmapala's essays and other writings discussed here were bundled in a 1965 collection: A. Dharmapala, *Return to Righteousness: A Collection of Speeches, Essays, and Letters of the Anagarika Dharmapala*, ed. A. W. Guruge (Anagarika Dharmapala Birth Centenary Committee, Ministry of Education and Cultural Affairs, Ceylon, 1965).

22. In a 1917 essay *The Nidanas or the Law of Dependent Causation* which focused on *paticca samuppada*, the fundamental Buddhist concept of dependent arising, Dharmapala also made connections to science:

> . . . it is instructive to note what the scientists, who are dealing with "radio-active process" have to say. Buddhism speaks of the continuous changes that are taking place without cessation in the atom, and it speaks of Sorrow, and suffering, depending on Change, and the uselessness of clinging to things which are momentarily changing. This non-clinging to things connected with the *skhandas* (aggregates), is beyond the comprehension of materialists, pantheists, deists, atheists, and monotheists.

23. In Dharmapala, *Return to Righteousness.*
24. DN 27.
25. *Abhidharma* (Sanskrit; *Abhidamma* in Pali) refers to the commentarial writings that accompany the *sutras* and *vinayas* in Buddhist canons. Theravada and Mahayana have their own sets of *Abhidharma* texts, with different sublineages having different commentarial texts as well. It is noteworthy that, again, Dharmapala has chosen to refer to the Sanskrit word instead of his native Pali, a trend pointed out by Lopez in *Buddhism & Science.*
26. Two additional noteworthy Dharmapala writings found in *Return to Righteousness* (1965) are "Evolution and Creation" (1917) and "Buddhism, Science and Christianity" (1924).
27. In addition to allegations such as associations between karmic imprints and human germ cells, much of Dharmapala's writing also reflected an influence of late nineteenth/early twentieth century race science and eugenics; this topic will be discussed later in Chapter 8.
28. In "Evolution from the Standpoint of Buddhism," Dharmapala made reference to a Buddhist belief in panspermia: the idea that life on earth came from an extraterrestrial source. Though certainly not a widespread belief in science today, panspermia hypotheses have been given serious consideration by prominent modern-era scientists such as Stephen Hawking, and space research programs today center much attention on the ability of different lifeforms and biomolecules to persist in space environments.
29. The Maha Bodhi Society is currently headquartered in Kolkutta, India, and has offices across many locations in Asia and other parts of the world. More information is available at mahabodhisocietyofindia.wordpress.com.
30. Bhikkuni Kusuma published a detailed autobiography in 2012: B. K. Kusuma, *Braving the Unknown Summit* (Colombo, Sri Lanka: Akna, 2012).
31. Kusuma, *Braving the Unknown Summit.*
32. Kusuma, *Braving the Unknown Summit.*
33. Kusuma, *Braving the Unknown Summit.*
34. She was joined by nine additional Sri Lankan women who would receive ordainment.
35. Kusuma also found support for her cause to restore the *bhikkuni* order to Sri Lanka in the diaries of Anagarika Dharmapala.
36. B. K. Kusuma, "How I Became a *Bhikkuni,*" *Sakyadhita Newsletter* 16, no. 2 (2008), 16–18.
37. D. R. Denver, P. C. Dolan, L. J. Wilhelm, W. Sung, J. I. Lucas-Lledó, D. K. Howe, S. C. Lewis, K. Okamoto, Michael Lynch Thomas, and C. F. Baer, "A Genome-Wide View of *Caenorhabditis elegans* Base-Substitution Mutation Processes," *Proceedings of the National Academy of Sciences USA* 106 (2009): 16310–16314.
38. Although some individuals who appear as biological females have one X and one Y chromosome, the situation typically found in men. There are also other natural deviations from the XX female/XY male paradigm that shapes our scientific thinking and societal definitions of biological sex. This topic will be covered in greater detail later in Chapter 8.

Chapter 5

1. Though the story of Laozi is a cherished central component of Taoism and broader Chinese history, some scholars doubt the historicity of Laozi as a historical figure, and instead view the *Tao Te Ching* as a compilation of writings from multiple unknown authors.
2. The "three teachings" are often described as peacefully coexisting in harmony throughout the history of China, though Chinese emperors usually favored one of the three teachings and discriminated against or at most tolerated the other two.
3. One *sutra* translated by An Shih-kao, important to the subsequent development of Zen, was titled in Chinese *Ta-an-pan-shou-i-ching* (*Sutra* on Concentration by Practicing Respiratory Exercises).
4. This record derived from an account written in the Book of Liang, compiled by the Chinese scholar Lao Shilian in 635 CE.
5. G. C. Godart, *Darwin, Dharma, and the Divine: Evolutionary Theory and Religion in Modern Japan* (Honolulu: University of Hawaii Press, 2017).
6. Shaku Soen's address was translated into English by his lay disciple, D. T. Suzuki, who would come to be one of the most prominent figures in the transmission of Zen Buddhism into Western culture.
7. K. Tsurumi, *The Minakata-Mandala: A Paradigm Change for the Future. UNESCO-UNU Joint Symposium on Science and Culture: A Common Path for the Future* (Tokyo: United Nations University, 1995).
8. Tsurumi, *The Minakata-Mandala*.
9. K. Minakata, "Constellations of the Far East," *Nature* 48 (1893), 541–543.
10. As outlined in Godart, *Darwin, Dharma, and the Divine*, a second Japanese biologist, Oka Asajiro, found limitations in the terminologies and approaches (e.g., the species concept) in biological taxonomy; he saw no rationale for objective boundaries between "species" and "variants." Asajiro further challenged the essential distinctions between germline and somatic cells in animal systems. Similar to Kusagumu, Asajiro studied enigmatic invertebrate organisms—"moss animals" (bryozoans) that displayed body plans that challenged the definition of an "individual."
11. Quotation from Godart, *Darwin, Dharma, and the Divine*.
12. N. Shin'ichi, "Minakata Kumagusu: The Meiji Polymath Who Broke the Mold," Nippon.com: Your Doorway to Japan, August 27, 2017.
13. T. Gill, "Reading Tsurumi Kazuko Reading Minakata Kumagusu," Tsurumi Kazuko Symposium, Kyoto Bunkyo University, Kyoto, Japan, June 19, 1999.
14. The Dalai Lama was not, however, the first Tibetan Buddhist to recognize Buddhism-science parallels. That distinction goes to Gendun Chopel, born in a northeastern region of Tibet in 1903 that traveled extensively around the world and wrote on the parallels between Buddhism and science in his travel journals, composed during 1934–1941. More details on this Tibetan traveler and pioneer of Buddhism-science intersections are provided in Wallace, *Buddhism & Science: Breaking New Ground*, and Lopez *Buddhism & Science*.

15. Dalai Lama XIV. *The Universe in a Single Atom: the Convergence of Science and Spirituality* (New York :Morgan Road Books, 2005).

16. J. A. Coyne, *Faith Versus Fact: Why Science and Religion Are Incompatible* (New York: Viking Press, 2015).

17. U. Goodenough, *The Sacred Depths of Nature* (New York: Oxford University Press, 1998).

18. Dalai Lama XIV, *The Universe in a Single Atom* (2005).

19. D. S. Wilson, *Does Altruism Exist? Culture, Genes, and the Welfare of Others* (New Haven, CT: Yale University Press, 2015).

20. J. Roughgarden, *The Genial Gene: Deconstructing Darwinian Selfishness* (Berkeley: University of California Press, 2009).

21. Godart, *Darwin, Dharma, and the Divine.*

22. A. Low, *The Origins of Human Nature: A Zen Buddhist Looks at Evolution* (Eastbourne, England: Sussex University Press, 2008).

23. R. Dawkins, *The Selfish Gene* (New York: Oxford University Press, 1976).

24. Dawkins, *The Selfish Gene.*

25. D. Noble, "Cardiac Action and Pacemaker Potentials Based on Hodgkin-Huxley Equations," *Nature* 188 (1960): 495–497.

26. D. Noble, *The Music of Life: Biology Beyond Genes* (New York: Oxford University Press, 2006).

27. The "genetic program" concept arose from the work of Francois Jacob and Jacques Monod, 1965 Nobel laureates whose reductionist-strategy research on the *Escherichia coli* lac operon system revealed important fundamental insights into the how genes are regulated (i.e., turned "on" and "off"), which subsequently came to be broadly characterized as the "genetic program." Mathieu Ricard, currently a monk of the Nyingma Tibetan tradition and author on diverse Buddhism-related topics, received his PhD in molecular genetics under the mentorship of Jacob before transitioning to the renunciate life.

28. Noble, *The Music of Life.*

29. D. Noble, "Convergence between Buddhism and Science: Systems Biology and the Concept of No-Self (*anatman*)," The 3rd World Conference on Buddhism and Science, Mahidol University, Nakhon Pathom, Thailand, 2010.

30. The modified prose, presented in Noble "Convergence between Buddhism and Science," was first published in D. Noble, "The Music of Life: A Systems Biology View of Buddhist Concepts of the Self/No Self," *Journal of the International Association of Buddhist Universities* 1 (2008): 1125–1139.

31. D. P. Barash, "The Ecologist as Zen Master," *The American Midland Naturalist* 89 (1973): 214–217.

32. D. P. Barash, *Buddhist Biology: Ancient Eastern Wisdom Meets Modern Western Science* (New York: Oxford University Press, 2013).

33. S. J. Gould, *Rocks of Ages: Science and Religion in the Fullness of Life* (New York: Ballantine Books, 2002).

34. Barash, *Buddhist Biology.*

35. T. N. Hanh, *Interbeing: Fourteen Guidelines for Engaged Buddhism* (Berkeley, CA: Parallax Press, 2003).

36. Similarities between Buddhist philosophy and Western philosophers such as David Hume and Immanuel Kant were also highlighted in Noble, "Convergence between Buddhism and Science" and have been thoroughly investigated in the humanities, alongside other postmodernism and other recent dialogues in the Western liberal arts traditions.

37. S. Brenner, "Life Sentences: Elementary Zenetics—Do or Dai," *Genome Biology* 3 (2002), comment1008.1–1008.2.

38. The Dalai Lama has pushed for the modernization of Gelug Tibetan Buddhist training and curriculum during his life, including the integration of scientific theories and practices, and allowing women into the monastic ranks.

39. F. W. Allendorf, "The Conservation Biologist as Zen Student," *Conservation Biology* 11 (1997): 1045–1046.

40. F. W. Allendorf, "Zen and Deep Evolution: The Optical Delusion of Separation," *Evolutionary Applications* 11 (2018): 1212–1218.

41. Coyne, *Faith Versus Fact.*

42. M. Kimura, *The Neutral Theory of Molecular Evolution* (Cambridge: Cambridge University Press, 1983).

43. A. Eisen and Y. Konchok, *The Enlightened Gene: Biology, Buddhism, and the Convergence That Explains the World* (Lebanon, NH; ForeEdge, 2018).

44. D. S. Wilson, *This View of Life: Completing the Darwinian Revolution* (New York: Pantheon Books, 2019).

45. Wilson, *This View of Life.*

46. M. Ricard and T. X. Thuan, *The Quantum and the Lotus: A Journey to the Frontiers Where Science and Buddhism Meet* (New York: Three Rivers, 2001).

47. V. Wallace and A. Wallace (organizers), Buddhism & Science Colloquium, Oxford University, March 4–5, 2010. Video footage of Denis Noble's introductory remarks available at https://www.youtube.com/watch?v=ujlfz_q5qcY

48. A pamphlet of Tibetan chants used to accompany the introduction and culmination of teachings at Maitripa College can be found here: https://fpmt.org/wp-content/uploads/education/EBP2006updates/pdf/prayers_for_teachings_bklta4.pdf

49. E. Alexander, *Proof of Heaven: A Neurosurgeon's Journey into the Afterlife* (New York: Simon & Schuster, 2012).

50. Dharmakirti's argument also underpinned the rationalization of rebirth set forth by Mathieu Ricard in *The Monk and the Philosopher.*

51. These three categories of phenomena are also sometimes presented with a fourth, preceding category of "false" or "illusory" phenomena—those that are not valid. Examples include mirages and notions of "self."

52. The ubiquitous occurrence of mitochondria in cells was first reported in 1890 by Altmann who initially referred to them as "bioblasts." More history of evidence for the existence of mitochondria is found in L. Ernster and G. Schatz, "Mitochondria: A Historical Review, *Journal of Cell Biology* 91 (1981): 227–255.

Chapter 6

1. This also generally includes rejection of the idea of "karmic" effects from one life to the next; i.e., the notion that actions in one's present life impact one's rebirth path in the next life. *Karma*, when more broadly considered as the Buddhist concept of mutual cause and effect as expressed in *pratityasamutpada*, is by contrast more generally acceptable to Western scientific minds.

2. The problems of "scientific materialism" are highlighted in the Introduction of B. A. Wallace, ed., *Buddhism & Science: Breaking New Ground* (New York: Columbia University Press, 2003).

3. H. H. Bauer, *Scientific Literacy and the Myth of the Scientific Method* (Urbana: University of Illinois Press, 1992).

4. Details of Craig Venter's life are provided in his autobiography: J. C. Venter, *A Life Decoded: My Genome—My Life* (New York: Penguin, 2007).

5. In the shotgun approach to genome sequencing developed by Venter and Celera Corp., very large amounts of DNA sequence reads are generated, to the extent that the total DNA sequenced is many times greater than the estimated total size of the target genome. With this approach, there is expected to be a large degree of overlap between the randomly generated sequences, and bioinformatic computer programs subsequently assemble the overlapping sequences into larger contiguous stretches of DNA. Ideally, complete chromosome sequences would be fully assembled with no gaps. This randomized approach contrasts to the previous approach, implemented by academic and government labs, whereby genome subsegments were first arranged in an orderly fashion through methodical molecular cloning techniques, followed by subsequent sequencing of the previously ordered DNA subsegments.

6. A transcript of then-President Clinton's speech announcing the completion of the human genome is available through the Clinton Presidential Materials Project of the National Archives and Records Administration: https://web.ornl.gov/sci/techresources/Human_Genome/project/clinton1.shtml

7. Both the public and privately funded human genome data became freely available shortly after the presidential announcement. Venter erroneously wagered that Celera Corp. could still make money by selling additional genetic information along with bioinformatics software packages required for analysis of the human genome information.

8. J. C. Venter et al., "Environmental Genome Shotgun Sequencing of the Sargasso Sea," *Science* 304 (2004): 66–74.

9. J. Shreeve, "Craig Venter's Epic Voyage to Redefine the Origin of the Species," *Wired*, August 1, 2004.

10. H. Andersen and B. Hepburn, "Scientific Method," in *The Stanford Encyclopedia of Philosophy* (Summer 2016 edition), ed. Edward N. Zalta, https://plato.stanford.edu/archives/sum2016/entries/scientific-method/

11. Andersen and Hepburn, "Scientific Method."

12. Parallels are apparent between Plato's deceptive nature of reality, as portrayed through the shadows in the cave allegory, and the deceptive nature of conventional reality according to Madhyamaka's Two Truths framework.

13. Andersen and Hepburn, "Scientific Method."

14. A. Falcon, "Aristotle on Causality," in *The Stanford Encyclopedia of Philosophy* (Spring 2019 edition), ed. Edward N. Zalta, https://plato.stanford.edu/archives/spr2019/entries/aristotle-causality/

15. Falcon, "Aristotle on Causality."

16. Andersen and Hepburn, "Scientific Method."

17. C. Hempel, *Aspects of Scientific Explanation and Other Essays in the Philosophy of Science* (New York: Free Press, 1965).

18. Andersen and Hepburn, "Scientific Method."

19. K. R. Popper, *The Logic of Scientific Discovery* (London: Routledge, 1959).

20. S. O. Hansson, "Science and Pseudo-Science," in *The Stanford Encyclopedia of Philosophy* (Summer 2017 edition), ed. Edward N. Zalta, https://plato.stanford.edu/archives/sum2017/entries/pseudo-science/

21. M. Pigliucci, "Pseudoscience," in *Encyclopedia of Philosophy and the Social Sciences*, ed. B. Kaldis (Thousand Oaks, CA: Sage, 2013), 766–769.

22. Hansson, "Science and Pseudo-Science."

23. M. Mahner, "Demarcating Science from Non-Science," in *Handbook of the Philosophy of Science: General Philosophy of Science—Focal Issues*, ed. T. Kuipers (Amsterdam: Elsevier BV, 2007), 515–575.

24. Though the Community of Knowledge Disciplines model presents science and reliable non-science scholarly traditions (i.e., the humanities and liberal arts), there exists a long-standing and ongoing commonly (though not universally) held sense of superiority among those in the sciences, that their epistemological methods offer fundamentally superior pathways to valid knowledge as compared to those of their colleagues in the humanities. Even within the sciences, there exists a common hierarchical framework whereby the physical sciences occupy a privileged status over the biological sciences, which in turn are afforded a higher status as compared to the psychological and social sciences. In fact, the latter two disciplines are often placed in liberal arts colleges at American universities, alongside non-science traditions such as philosophy and religion.

25. Hansson, "Science and Pseudo-Science."

26. K. R. Popper, *Conjectures and Refutations: The Growth of Scientific Knowledge* (New York: Basic Books, 2002).

27. Hansson, "Science and Pseudo-Science."

28. M. Ruse, "Is Evolutionary Biology a Different Kind of Science?," *Aquinas* 43 (2000): 251–282.

29. Hansson, "Science and Pseudo-Science."

30. A. A. Derkson, "The Seven Sins of Pseudoscience," *Journal for General Philosophy of Science* 24 (1993): 17–42.

31. H. Frielander, *The Origins of Nazi Genocide: From Euthanasia to the Final Solution* (Chapel Hill: University of North Carolina Press, 2000).

32. Though human rebirth is a fundamental belief central to almost all ancient and modern forms of Buddhism, one of the early eighteen schools that evolved in India during the tradition's early diversification (from ~300 BCE to 500 CE) is said to have not held human rebirth as a core doctrine (J. Blumenthal, personal communication). Modern "secular Buddhism" movements, largely motivated by the works of Stephen Batchelor, propose paths that are free from rebirth concepts but still identify themselves as "Buddhist" traditions.

Chapter 7

1. Search performed on March 23, 2021, at https://www.ncbi.nlm.nih.gov/pubmed
2. The conceptualizations of *pratityasamutpada* by Piyadissi Thera (covered in Chapter 2), Nagarjuna (covered in Chapter 3), and Fazan (covered in Chapter 4) contributed to the development and falsification strategy applied to H_s3.
3. I. Hargittai, "Linus Pauling's Question for the Structure of Proteins," *Structural Chemistry* 21 (2010): 1–7.
4. O. T. Avery, C. M. MacLeod, and M. McCarty, "Studies on the Chemical Nature of the Substance Inducing Transformation of Pneumococcal Types," *Journal of Experimental Medicine* 79 (1944): 137–158.
5. "Linus Pauling and the Race for DNA," Special Collections and Archives Research Center, Oregon State University, 2015, http://scarc.library.oregonstate.edu/coll/pauling/dna/narrative/page5.html
6. L. Pauling and R. B. Corey, "A Proposed Structure for the Nucleic Acids," *Proceedings of the National Academy of Sciences USA* 39 (1953): 84–97.
7. J. D. Watson and F. H. C. Cricke, "Molecular Structure of Nucleic Acids: A Structure for Deoxyribose Nucleic Acid," *Nature* 171 (1953): 737–738.
8. "Linus Pauling and the Race for DNA," Special Collections and Archives Research Center, Oregon State University, 2015, http://scarc.library.oregonstate.edu/coll/pauling/dna/narrative/page33.html
9. One cycle in a typical PCR involves three temperature steps: denaturation (usually 94°C–98°C, where the target DNA becomes single-stranded), annealing (usually 50°C–70°C, where short single-stranded DNA "primers" added to the reaction bind to complementary regions on the target DNA), and extension (usually 72°C, where the DNA polymerases create new copies of DNA using the original target DNA as a template). Most PCRs involve twenty-five to thirty-five cycles, with billions of copies of the target DNA segment resulting once all cycles are completed. Multiple scientific manufacturers produce machines, called thermocyclers, to apply these temperature cycle regimens to diverse PCR applications in biology labs.
10. K. Mullis, *Dancing Naked in the Mind Field* (New York: Random House, 2000).
11. G. Bonner and A. M. Klibanov, "Structural Stability of DNA in Nonaqueous Solvents," *Biotechnology and Bioengineering* 69 (2000): 339–344.
12. V. Sertic and N. Bulgakov, "Classification et identification des typhi-phage," *C. R. Soc. Biol. Paris* 119 (1935): 1270–1272.

13. R. L. Sinsheimer, "A Single-Stranded Deoxyribonucleic Acid from Bacteriophage φX174," *Journal of Molecular Biology* 1 (1959): 43–53.

14. F. Sanger, S. Nicklen, and A. R. Coulson, "DNA Sequencing with Chain-Terminating Inhibitors," *Proceedings of the National Academy of Sciences USA* 74 (1977): 5463–5467.

15. H. O. Smith, C. A. Huchinson III, C. Pfannkoch, and J. C. Venter, "Generating a Synthetic Genome by Whole Genome Assembly: φX174 Bacteriophage from Synthetic Oligonucleotides," *Proceedings of the National Academy of Sciences USA* 100 (2003): 15440–15445.

16. S. Modrow, D. Falke, U. Truyen, and H. Schatzl, *Molecular Virology* (Berlin: Springer-Verlag, 2013).

17. M. Yoshida, T. Mochizuki, S.-I. Uruyama, Y. Yoshida-Takashima, S. Nishi, M. Hirai, H. Nomaki, Y. Takaki, T. Nonoura, and K. Takai "Quantitative Viral Community DNA Analysis Reveals the Dominance of Single-Stranded DNA Viruses in Offshore Upper Bathyal Sediment from Tohoku, Japan," *Frontiers in Microbiology* 9 (2018): 75.

18. C. M. Gedik, S. P. Boyle, S. G. Wood, N. J. Vaughan, and A. R. Collins, "Oxidative Stress in Humans: Validation of Biomarkers of DNA Damage," *Carcinogenesis* 23 (2002): 1441–1446.

19. C. Park, J. D. Rosenblat, E. Brietzke, Z. Pan, Y. Lee, B. Cao, H. Zuckerman, A. Kalantarova, and R. S. McIntyre (2019), "Stress, Epigenetics and Depression: A Systematic Review," *Neuroscience and Biobehavioral Reviews* 102 (2019): 139–152.

20. A. Unnikrishnan, W. F. Freeman, J.'Jackson, J. D. Wren, H. Porter, and A. Richardson, "The Role of DNA Methylation in Epigenetics of Aging," *Pharmacology & Therapeutics* 195 (2019): 172–185.

21. M. G. Gall and T. H. Bestor, "Eukaryotic Cytosine Methyltransferases," *Annual Review of Biochemistry* 74 (2005): 481–514.

22. E. Li, T. H. Bestor, and R. Jaenisch, "Targeted Mutation of the DNA Methyltransferase Gene Results in Embryonic Lethality," *Cell* 69 (1992): 915–926.

23. M. Okano, D. W. Bell, D. A. Haber, and E. Li, "DNA Methyltransferases Dnmt3a and Dnmt3b Are Essential for De Novo Methylation and Mammalian Development," *Cell* 99 (1999): 247–257.

24. S. Kumar, V. Chinnusamy, and T. Mohapatra, "Epigenetics of Modified DNA Bases: 5-Methylcytosine and Beyond," *Frontiers in Genetics* 9 (2018): 640.

25. T. Tong, S. Chen, L. Wang, Y. Tang, J. Y. Ryu, S. Jiang, X. Wu, C. Chen, J. Luo, Z. Deng, Z. Li, S. Y. Li, and S. Chen. "Occurrence, Evolution, and Functions of DNA Phosphorothioate Epigenetics in Bacteria," *Proceedings of the National Academy of Sciences USA* 115 (2018): E2988–E2996.

26. E. M. Bradbury, W. C. Price, G. R. Wilkinson, and G. Zubay, "Polarized Infrared Studies of Nucleoproteins. II. Nucleohistone," *Journal of Molecular Biology* 4 (1962): 50–60.

27. S. Brenner, F. Jacob, and M. Meselson, "An Unstable Intermediate Carrying Information from Genes to Ribosomes for Protein Synthesis," *Nature* 190 (1961): 576–580.

28. F. H. Crick, L. Barnett, S. Brenner, and R. J. Watts-Tobin, "General Nature of the Genetic Code for Proteins," *Nature* 192 (1961): 1227–1232.

29. E. V. Koonin and A. S. Novozhilov, "Origin and Evolution of the Genetic Code: The Universal Enigma," *IUBMB Life* 61 (2009): 99–111.

30. B. G. Barrell, A. T. Bankier, and J. Drouin, "A Different Genetic Code in Human Mitochondria," *Nature* 282 (1979): 189–194.

31. D. V. Lavrov and W. Pett, "Animal Mitochondrial DNA As We Do Not Know It: Mt-Genome Organization and Evolution in Nonbilaterian Lineages," *Genome Biology and Evolution* 8 (2016): 2896–2913.

32. J. Ling, P. O'Donoghue, and D. Soll, "Genetic Code Flexibility in Micro-organisms: Novel Mechanisms and Impact on Physiology," *Nature Reviews Microbiology* 13 (2015): 707–721.

33. J. W. Chin, "Expanding and Reprogramming the Genetic Code," *Nature* 550 (2017): 53–60.

34. Careful consideration of other molecular processes associated with biological information processing systems, such as translation and RNA splicing, yields similar results.

35. The fitness advantage associated with a mutation in future generations depends upon other genetic considerations such as dominance relationships (i.e., whether that mutation is in a homozygous or heterozygous state, if in a diploid organism) and epistasis (i.e., the interactions of the gene's product with gene products).

36. J. Lederberg and E. M. Lederberg, "Replica Plating and Indirect Selection of Bacterial Mutants," *Journal of Bacteriology* 63 (1952): 399–406.

37. S. E. Luria and M. Delbruck, "Mutations of Bacteria from Virus Sensitivity to Virus Resistance," *Genetics* 28 (1943): 491–511.

38. J. Ollila, I. Lappalainen, and M. Vihinen, "Sequence Specificity in CpG Mutation Hotspots," *FEBS Letters* 396 (1996): 119–122.

39. K. A. Eckert and S. E. Hile, "Every Microsatellite Is Different: Intrinsic DNA Features Dictate Mutagenesis of Common Microsatellites Present in the Human Genome," *Molecular Carconigenesis* 48 (2009): 379–388.

40. I. Martincorena, A. S. Seshasayee, and L. M. Luscombe, "Evidence of Non-random Mutation Rates Suggests an Evolutionary Risk Management Strategy," *Nature* 485 (2012): 95–98.

41. B. Schuster-Bockler and B. Lehner, "Chromatin Organization Is a Major Influence on Regional Mutation Rates in Human Cancer Cells," *Nature* 488 (2012): 504–507.

42. The blurring of lines between solute and solvent especially resonates with Vietnamese monk, author, and peace activist Thich Nhat Hanh's "interbeing" conceptualization of *pratitysamutpada*. The solutes and the solvents "inter-are."

Chapter 8

1. R. O. Wood and V. Orel, *Genetic Prehistory in Selective Breeding* (New York: Oxford University Press, 2001).

2. F. C. Richter, "Remembering Johann Gregor Mendel: A Human, a Catholic Priest, an Augustinian Monk, and Abbot," *Molecular Genetics & Genomic Medicine* 3 (2015): 483–485.

3. M. De Castro, "Johann Gregor Mendel: Paragon of Experimental Science," *Molecular Genetics & Genomic Medicine* 4 (2016): 3–8.

4. J. M. Opitz and D. W. Bianchi, "Mendel: Morphologist and Mathematician Founder of Genetics—To Begin a Celebration of the 2015 Sesquicentennial of Mendel's Presentation in 1865 of His Versuche über Pflanzenhybriden," *Molecular Genetics & Genomic Medicine* 3 (2015): 1–7.

5. In more technical genetic terms, because the genes were on separate chromosomes, they underwent independent assortment—this in fact was one major rule of genetics that arose from Mendel's pea plant experiments. When two genes are located very close to one another, the segregation patterns of the two genes are not independent— they are "linked."

6. De Castro, "Johann Gregor Mendel."

7. A. H. Sturtevant, *A Short History of Genetics* (New York: Harper and Row, 1965).

8. R. A. Fisher, "Has Mendel's Work Been Rediscovered?" *Annals of Science* 1 (1936): 115–137.

9. De Castro, "Johann Gregor Mendel."

10. A. Hershey and M. Chase, "Independent Functions of Viral Protein and Nucleic Acid in Growth of Bacteriophage," *Journal of General Physiology* 36 (1952): 39–56.

11. W. Hennig, *Phylogenetic Systematics* (Urbana: University of Illinois Press, 1966).

12. V. Savolainen, R. S. Cowan, A. P. Vogler, G. K. Roderick, and R. Lane, "Towards Writing the Encyclopaedia of Life: An Introduction to DNA Barcoding," *Philosophical Transactions of the Royal Society B* 360 (2005): 1805–1811.

13. K. Bohmann, A. Evans, M. Gilbert, P. Thomas, G. R. Carvalho, S. Creer, M. Knapp, D. W. Yu, and M. de Bruyn, "Environmental DNA for Wildlife Biology and Biodiversity Monitoring," *Trends in Ecology & Evolution* 29 (2014): 358–367.

14. It is also noteworthy to the topic of this book that Lamar mentions meditation later in the song, though in a context inconsistent with the Buddha's teachings: "I don't con- template, I meditate, then off your fucking head."

15. Each chromosome in humans is composed of DNA and associated proteins, the latter involved in mediating diverse DNA-associated functions such as transcriptional reg- ulation and chromosomal segregation during cell division processes.

16. Boveri and Sutton each independently formulated the chromosomal theory of inheritance.

17. P. C. Watts, K. R. Buley, S. Sanderson, W. Boardman, C. Ciofi, and R. Gibson, "Parthenogenesis in Komodo Dragons," *Nature* 444 (2006): 1021–1022.

18. D. M. Chapman, M. S. Shivji, E. Louis, J. Sommer, H. Fletcher, and P . A. Prodohl, "Virgin Birth in a Hammerhead Shark," *Biology Letters* 3 (2007): 425–427.

19. C. Moritz, "Parthenogenesis in the Endemic Australian Lizard *Heteronotia binoei* (Gekkonidae)," *Science* 220 (1983): 735–737.

20. J. Volmink and B. Marais, "HIV: Mother to Child Transmission," *Clinical Evidence* 2 (2008): 909.

21. N. Vu Lam, P. B. Gotsch, and R. C. Langan, "Caring for Pregnant Women and Newborns with Hepatitis B or C," *American Family Physician* 82 (2010): 1225–1229.

22. Information collected on March 24, 2021, from https://www.cdc.gov/nchs/nvss/vsrr/covid19/index.htm

23. W. M. Liu, R. T. K. Pang, P. C. Chiu, B. P. C. Wong, K. Lao, K.-F. Lee, and W. S. B. Yeung. "Sperm-Borne MicroRNA-34c Is Required for the First Cleavage Division in Mouse," *Proceedings of the National Academy of Sciences USA* 109 (2012): 490–494.

24. S. Yuan, A. Schuster, C. Tang, T. Yu, N. Ortogero, J. Bao, H. Zheng, and W. Yan. "Sperm-Borne miRNAs and Endo-siRNAs Are Important for Fertilization and Preimplantation Embryonic Development," *Development* 143 (2016): 635–647.

25. Q. Chen, W. Yan, and E. Duan, "Epigenetic Inheritance of Acquired Traits through Sperm RNAs and Sperm RNA Modifications," *Nature Reviews Genetics* 17 (2016): 733–743.

26. A. B. Rodgers, C. P. Morgan, N. A. Leu, and T. L. Bale, "Transgenerational Epigenetic Programming via Sperm microRNA Recapitulates Effects of Paternal Stress," *Proceedings of the National Academy of Sciences USA* 112 (2015): 13699–13704.

27. Q. Chen, M. Yan, Z. Cao, X. Li, Y. Zhang, J. Shi, G. Feng, H. Peng, X. Zhang, Y. Zhang, J. Qian, E. Duan, Q. Zhai, and Q. Zhou. "Sperm tsRNAs Contribute to Inter-generational Inheritance of an Acquired Metabolic Disorder," *Science* 351 (2016): 397–400.

28. E. L. Greer and Y. Shi, "Histone Methylation: A Dynamic Mark in Health, Disease and Inheritance," *Nature Reviews Genetics* 13 (2012): 343–357.

29. R. T. Johnson, "Prion Diseases," *The Lancet Neurology* 4 (2005): 635–642.

30. J. O. Hoskin, L. G. Kiloh, and J. E. Cawte, "Epilepsy and Guria: The Shaking Syndromes of New Guinea," *Social Science & Medicine* 3 (1969): 39–48.

31. Z. H. Harvey, Y. Chen, and D. F. Jarosz, "Protein-Based Inheritance: Epigenetics beyond the Chromosome," *Molecular Cell* 69 (2017): 195–202.

32. C. R. Woese, O. Kandler, and M. L. Wheelis, "Towards a Natural System of Organisms: Proposal for the Domains Archaea, Bacteria, and Eucarya," *Proceedings of the National Academy of Sciences USA* 87 (1990): 4576–4579.

33. J. R. Brown and W. F. Doolittle, "Archaea and the Prokaryote-to-Eukaryote Transition," *Microbiology and Molecular Biology Reviews* 61 (1997): 456–502.

34. W. F. Doolittle, "Phylogenetic Classification and the Universal Tree," *Science* 284 (1999): 2124–2128.

35. V. Daubin and G. J. Szollosi, "Horizontal Gene Transfer and the History of Life," *Cold Spring Harbor Perspectives in Biology* 8 (2016): a018036.

36. A. Lerner, T. Matthias, and R. Aminov, "Potential Effects of Horizontal Exchange in the Human Gut," *Frontiers in Immunology* 8 (2017): 1630.

37. S. M. Soucy, J. Huang, and J. P. Gogarten, "Horizontal Gene Transfer: Building the Web of Life," *Nature Reviews Genetics* 16 (2015): 472–482.

38. J. C. Dunning Hotopp, "Horizontal Gene Transfers between Bacteria and Animals," *Trends in Genetics* 27 (2011): 157–163.

39. N. Kondo, N. Nikoh, N. Ijichi, M. Shimada, and T. Fukatsu. "Genome Fragment of *Wolbachia* Endosymbiont Transferred to X Chromosome of Host Insect," *Proceedings of the National Academy of Sciences USA* 99 (2002): 14280–14285.

40. J. C. D. Hottop, M. E. Clark, D. C. S. G. Oliveira, J. M. Foster, P. Fischer, M. C. M. Torres, J. D. Giebel, N. Kumar, N. Ishmael, S. Wang, J. Ingram, R. V. Nene, J. Shepard, J. Tomkins, S.

Richard. D. J. Spiro, E. Ghedin, B. E. Slatko, H. Tettelin, and J. H. Werren. "Widespread Lateral Gene Transfer from Intracellular Bacteria to Multicellular Eukaryotes," *Science* 317 (2007): 1753–1756.

41. S. N. McNulty, J. M. Foster, M. Mitreva, J. C. D. Hotopp, J. Martin, K. Fischer, B. Wu, P. J. Davis, S. Kumar, N. W. Brattig, B. E. Slatko, G. J. Weil, and P. U. Fischer. "Endosymbiont DNA in Endosymbiont-Free Filarial Nematodes Indicates Ancient Horizontal Gene Transfer," *PLoS One* 5 (2010): e11029.

42. E. G. J. Danchin, M. N. Rosso, P. Vieira, J. de Almeida-Engler, P. M. Coutinho, B. Henrissat, and P. Abaad. "Multiple Lateral Gene Transfers and Duplications Have Promoted Plant Parasitism Ability in Nematodes," *Proceedings of the National Academy of Sciences USA* 107 (2010): 17651–17656.

43. Y. M. D. Lo, R. W. K. Chiu, K. C. A. Chan, T. Y. Leung, T. K. Lau, and Y. M. D. Lo. "Microfluidics Digital PCR Reveals a Higher Than Expected Fraction of Fetal DNA in Maternal Plasma," *Clinical Chemistry* 54 (1998): 1664–1672.

44. Y. M. D. Lo, K. C. A. Chan, H. Sun, E. Z. Chen, P. Jiang, M. F. M. Lun, Y. W. Zheng, T. Y. Leung, T. K. Lau, C. R. Cantor, and R. W. K. Chiu. "Maternal Plasma DNA Sequencing Reveals the Genome-Wide Genetic and Mutational Profile of the Fetus," *Science Translational Medicine* 61 (2010): 61ra91.

45. D. W. Bianchi, G. K. Zickwolf, G. J. Weil, S. Sylvester, and M. A. DeMaria, "Male Fetal Progenitor Cells Persist in the Maternal Blood for as Long as 27 Years Postpartum," *Proceedings of the National Academy of Sciences USA* 93 (1996): 705–708.

46. W. F. N. Chan, C. Gurnot, T. J. Montine, J. A. Sonnen, K. A. Guthrie, and J. Lee Nelson, *PLoS One* 7 (2012): e45592.

47. B. Milholland, X. Dong, L. Zhang, X. Hao, Y. Suh, and J. Vijg, "Differences between Germline and Somatic Mutation Rates in Humans and Mice," *Nature Communications* 8 (2017): 15183.

48. M. J. McConnell, M. R. Lindberg, K. J. Brennand, J. C. Piper, T. Voet, C. Cowing-Zitron, S. Shumilina, R. S. Laskin, J. R. Vermeesch, I. M. Hall, and F. H. Gage. "Mosaic Copy Number Variation in Human Neurons," *Science* 342 (2013): 632–637.

49. T. E. Starzl and A. J. Demetris, "Transplantation Tolerance, Microchimerism, and the Two-Way Paradigm," *Theoretical Medical Bioethics* 19 (1998): 441–455.

50. M. S. Kruskall, T. H. Lee, F. S. Assmann, M. Laycock, L. A. Kalish, M. M. Lederman, M. P. Busch, and Viral Activation Transfusion Study Group. "Survival of Transfused Donor White Cells in HIV-Infected Recipients," *Blood* 98 (2001): 272–279.

51. H. R. Maturana and F. J. Valera, *Autopoiesis and Cognition: The Realization of the Living* (Boston: Reidel, 1980).

52. F. J. Valera, "Intimate Distances—Fragments for a Phenomenology of Organ Transplantation," *Journal of Consciousness Studies* 8 (2001): 259–271.

53. C. Darwin, *The Descent of Man, and Selection in Relation to Sex* (London: John Murray, 1871), 71.

54. Grayson Perry, *The Descent of Man* (London: Allan Lane, 2016).

55. N. Valenzuela and V. A. Lance, *Temperature-Dependent Sex Determination in Vertebrates* (Washington, DC: Smithsonian Books, 2004).

56. N. J. Mitchell, M. R. Kearney, N. J. Nelson, and W. P. Porter, "Predicting the Fate of a Living Fossil: How Will Global Warming Affect Sex Determination and Hatching Phenology in Tuatara?" *Proceedings of the Royal Society* B 275 (2008): 2185–2193.

57. F. J. Janzen and P. C. Phillips, "Exploring the Evolution of Environmental Sex Determination, Especially in Reptiles," *Journal of Evolutionary Biology* 19 (2006): 1775–1784.

58. J. Nielsen and M. Wohlert, "Sex Chromosome Abnormalities Found among 34,910 Newborn Children: Results from a 13-Year Incidence Study in Arhus, Denmark," *Birth Defects Original Article Series* 26 (1990): 209–223.

59. Some individuals diagnosed with Turner syndrome have one intact X chromosome and a second X chromosome that is highly truncated, missing substantial amounts of DNA.

60. R. M. Baxter and E. Vilain, "Translational Genetics for Diagnosis of Human Disorders of Sex Development," *Annual Review of Genomics and Human Genetics* 4 (2013): 371–392.

61. M. Blackless, A. Charuvastra, A. Derryck, A. Fausto-Sterling, K. Lauzanne, and E. Lee, "How Sexually Dimorphic Are We? Review and Synthesis," *American Journal of Human Biology* 12 (2000): 151–166.

62. P. Pagonis, "The Son They Never Had," *Narrative Inquiry in Bioethics* 5 (2015): 103–106.

63. E. L. Green, K. Benner, and R. Pear, "'Transgender' Could Be Defined out of Existence under Trump Administration," *New York Times*, October 21, 2018.

64. D. Reich, "How Genetics Is Changing Our Understanding of 'Race,'" *New York Times*, March 23, 2018.

65. J. D. Eisenback and H. H. Triantaphyllou, "Root-Knot Nematodes: *Meloidogyne* Species and Races," in *Manual of Agricultural Nematology*, ed. W. R. Nickle, 191–274 (New York: Marcell Dekker, 1991).

66. In positive eugenics, individuals with desirable traits (e.g., high intelligence) were encouraged to mate with others with positive traits. In negative eugenics, individuals with negative traits (e.g., feeblemindedness, sexual deviancy) were discouraged from reproducing.

67. S. J. Gould, *The Mismeasure of Man* (New York: Norton, 1981).

68. R. Lewontin, "The Apportionment of Human Diversity," *Evolutionary Biology* 6 (1972): 391–398.

69. https://www.ted.com/talks/svante_paabo_dna_clues_to_our_inner_neanderthal?language=en#t-31113

70. During the late 2010s, genomic analyses suggested that Neanderthal DNA was present in the genomes of people of European and Asian origins, but not those from Africa. A 2020 study by Princeton geneticist Joshua Akey and colleagues, however, suggested that African individuals also harbor Neanderthal DNA. Chen et al., "Identifying and Interpreting Apparent Neanderthal Ancestry in African Individuals," *Cell* (2020).

Chapter 9

1. J. Shreeve, "Craig Venter's Epic Voyage to Redefine the Origin of the Species," *Wired*, August 1, 2004.

2. *American Masters: Decoding Watson*, directed by Mark Mannucci, starring James Watson, Public Broadcasting Service, 2019.

3. M. Kimura, *The Neutral Theory of Molecular Evolution* (Cambridge: Cambridge University Press, 1983).

4. J. F. Crow, "Memories of Moto," *Theoretical Population Biology* 49 (1996): 122–127.

5. I. Yanai and M. Lercher, "A Hypothesis Is a Liability," *Genome Biology* 21 (2020): 231.

6. Chandrakirti, Padmakara Translation Group, transl., *Introduction to the Middle Way* (Boston: Shambhala, 2005).

7. D. Arnold, "How to Do Things with Candrakirti: A Comparative Study in Anti-Skepticism," *Philosophy East and West* 51 (2001): 247–279.

8. T. H. Hanh, *Interbeing: Fourteen Guidelines for Engaged Buddhism* (Berkeley, CA: Parallax Press, 1987).

9. K. Crenshaw, "The Urgency of Intersectionality," https://www.ted.com/talks/kimberle_crenshaw_the_urgency_of_intersectionality?language=en

10. A. Fleming, "On the Antibacterial Action of Cultures of a Penicillium, with Special Reference to Their Use in the Isolation of *B. influenza*," *British Journal of Experimental Pathology* 10 (1929): 226–236.

11. K. Dunbar and J. Fuselang, "Scientific Thinking and Reasoning," in *The Cambridge Handbook of Thinking and Reasoning*, ed. K. J. Holyoak and R. G. Morrison (New York: Cambridge University Press, 2005), 705–725.

12. A. A. Bauermeister, "Serendipity and the Cerebral Localization of Pleasure," *Neoplasma* 23 (1976): 259–263.

13. Bhikku Bodhi, transl. *Numerical Discourses of the Buddha: A Translation of the Anguttara Nikaya* (Somerville, MA: Wisdom Publications, 2012).

14. A. Brahm, *Mindfulness, Bliss, and Beyond* (Somerville, MA: Wisdom Publications, 2006).

15. An apt summary of the health benefits of mindfulness meditation are shared on the National Institutes of Health National Center for Complementary and Integrative Health website focused on this topic: https://www.nccih.nih.gov/health/meditation-in-depth

16. R. J. Odzer, "Tilopa's Six Nails," *Tricycle* (Spring 2018).

17. T. K. Desta and T. Mulugeta, "Living with COVID-19-Triggered Pseudoscience and Conspiracies," *International Journal of Public Health* June 29 (2020): 1–2.

Chapter 10

1. K. Joyce, "The Trouble with the Christian Adoption Movement," *The New Republic*, January 11, 2016.

2. K. Joyce, *The Child Catchers: Rescue, Trafficking, and the New Gospel of Adoption* (New York: PublicAffairs, 2013).

3. Joyce, "The Trouble with the Christian Adoption Movement."

4. W. D. Hamilton, "The Genetical Evolution of Social Behavior, I and II," *Journal of Theoretical Biology* 7 (1964): 1–52.

5. J. Maynard Smith, "Group Selection and Kin Selection," *Nature* 201 (1964): 1145–1147.

6. R. Dawkins, "Burying the Vehicle Commentary on Wilson & Sober: Group Selection," *Behavioral and Brain Sciences* 17 (1994): 616–617.

7. D. S. Wilson and E. Sober, "Reintroducing Group Selection to the Human Behavioral Sciences," *Behavioral and Brain Sciences* 17 (1994): 585–654.

8. M. A. Nowak, C. E. Tarnita, and E. O. Wilson, "The Evolution of Eusociality," *Nature* 466 (2010): 1057–1062.

9. P. Abbot, J. Abe, J. Alcock, S. Alizon, J. A. C. Alpedrinha, M. Andersson, J.-B. Andre, M. van Baalen, F. Balloux, S. Balshine, N. Barton, L. W. Beukeboom, J. M. Biernaskie, T. Bilde, G. Borgia, M. Breed, S. Brown, R. Bshary, A. Buckling, N. T. Burley, M. N. Burton-Chellew, M. A. Cant, M. Chapuisat, E. L. Charnov, T. Clutton-Brock, A. Cockburn, B. J. Cole, N. Colegrave, L. Cosmides, I. D. Couzin, J. A. Coyne, S. Creel, B. Crespi, R. L. Curry, S. R. X. Dall, T. Day, J. L. Dickinson, L. A. Dugatkin, C. El Mouden, S. T. Emlen, J. Evans, R. Ferriere, J. Field, S. Foitzik, K. Foster, W. A. Foster, C. W. Fox, J. Gadau, S. Gandon, A. Gardner, M. G. Gardner, T. Getty, M. A. D. Goodisman, A. Grafen, R. Grosberg, C. M. Grozinger, P.-H. Gouyon, D. Gwynne, P. H. Harvey, B. J. Hatchwell, J. Heinze, H. Helantera, K. R. Helms, K. Hill, N. Jiricny, R. A. Johnstone, A. Kacelnik, E. T. Kiers, H. Kokko, J. Komdeur, J. Korb, D. Kronauer, R. Kümmerli, L. Lehmann, T. A. Linksvayer, S. Lion, B. Lyon, J. A. R. Marshall, R. McElreath, Y. Michalakis, R. E. Michod, D. Mock, T. Monnin, R. Montgomerie, A. J. Moore, U. G. Mueller, R. Noë, S. Okasha, P. Pamilo, G. A. Parker, J. S. Pedersen, I. Pen, D. Pfennig, D. C. Queller, D. J. Rankin, S. E. Reece, H. K. Reeve, M. Reuter, G. Roberts, S. K. A. Robson, D. Roze, F. Rousset, O. Rueppell, J. L. Sachs, L. Santorelli, P. Schmid-Hempel, M. P. Schwarz, T. Scott-Phillips, J. Shellmann-Sherman, P. W. Sherman, D. M. Shuker, J. Smith, J. C. Spagna, B. Strassmann, A. V. Suarez, L. Sundström, M. Taborsky, P. Taylor, G. Thompson, J. Tooby, N. D. Tsutsui, K. Tsuji, S. Turillazzi, F. Úbeda, E. L. Vargo, B. Voelkl, T. Wenseleers, S. A. West, M. J. West-Eberhard, D. F. Westneat, D. C. Wiernasz, G. Wild, R. Wrangham, A. J. Young, D. W. Zeh, J. A. Zeh & A. Zink "Inclusive Fitness Theory and Eusociality," *Nature* 471 (2011): E1–E4.

10. W. FitzPatrick, "Morality and Evolutionary Biology," in *The Stanford Encyclopedia of Philosophy*, ed. Edward N. Zalta, https://plato.stanford.edu/entries/morality-biology/

11. R. Dawkins, *The God Delusion* (New York: Houghton Mifflin, 2006).

12. FitzPatrick, "Morality and Evolutionary Biology."

13. M. Lynch, *The Origins of Genome Architecture* (Sunderland, MA: Sinauer Associates, 2007).

14. N. Carbone, "Kony 2012/Arts & Entertainment/TIME.com," December 4, 2012.

15. A. Van Woudenberg, "How to Catch Joseph Kony/Human Rights Watch," Human Rights Watch, March 9, 2012.

16. Comments made at the 2008 Academy of Achievement Summit: http://www.achievement.org/summit/2008

17. T. Cole, "The White Savior Industrial Complex," *The Atlantic*, March 22, 2012.

Glossary

anatman (Sanskrit; *anatta* in Pali) commonly translated as non-self.

anitya (Sanskrit; *anicca* in Pali) commonly translated as impermanence.

arhat (Sanskrit; *arahant* in Pali) an individual who has achieved enlightenment (in Theravada Buddhism) or made progress along the path to enlightenment (in Mahayana Buddhism).

bhiksu (Sanskrit; *bhikku* in Pali) an ordained Buddhist monk.

bhiksuni (Sanskrit; *bhikkuni* in Pali) an ordained Buddhist nun.

Bodhi (Sanskrit and Pali) wisdom, enlightenment.

bodhicitta (Sanskrit) spontaneous and universal compassion for all beings.

bodhisattva (Sanskrit; *bodhisatta* in Pali) beings on the path to becoming Buddhas (in Theravada) or beings motivated by *bodhicitta* to achieve enlightenment to liberate all beings from suffering (in Mahayana).

Buddha a being who has achieved complete enlightenment, such as Siddhartha Gautama.

conventional truth valid reality as experienced by beings who properly understand the nature of the universe as described in Madhyamaka philosophy.

dharma (Sanskrit; *dhamma* in Pali) the Buddha's teachings, or phenomena in the natural world as properly understood through Buddhist teachings.

DNA deoxyribonucleic acid; a biomolecule that often serves as heritable genetic material.

DNA sequence the linear order of A's, C's, G's, and T's on a particular single DNA strand.

duhkha (Sanskrit; *dukkha* in Pali) suffering, dissatisfaction, or uneasiness; also simply refers to the lived experience.

epigenetics the study of heritable changes that do not involve changes in the DNA sequence.

genome the entire composition of DNA inside a cell or organism.

Hinayana (Sanskrit) commonly translated as "lesser vehicle"; refers to early schools of Buddhism considered to have inferior or incomplete Buddhist teachings as compared to Mahayana.

hypothesis a proposed explanation for some for some process or phenomenon.

intersectionality an identity framing concept that considers an individual's unique experience as an intersection of many identity-influencing categories such as race, ethnicity, gender, ability, religion, and so on.

intersex a term to describe individuals with sex characteristics that do not fit the binary "male and female" norms of society.

isotope one of two or more forms of an element that differ in their subatomic composition and atomic weight.

Madhyamaka (Sanskrit) a Mahayana Buddhist school of philosophy, also known as Middle Way philosophy, that originated in India and further developed in Tibet.

Mahayana (Sanskrit) commonly translated as "greater vehicle"; one of two widely recognized branches of Buddhism (along with Theravada) practiced in a variety of places such as China, Tibet, Japan, Korea, and Vietnam.

mutation a DNA sequence change.

nirvana (Sanskrit; *nibbana* in Pali), a transcendent state where the cycle of birth and death ends and all suffering is quenched.

PCR polymerase chain reaction; a molecular biology method whereby billions of target DNA molecules can be created in a tube, starting from a small amount of input DNA.

phage a virus that infects bacteria.

pratityasamutpada (Sanskrit; *paticca samuppada* in Pali) commonly translated as dependent arising or dependent origination; the Buddha's cause-and-effect framework for interconnectivity and how the universe functions.

prion misfolded proteins capable of transmitting their misfolded shape on to other proteins.

protein a biomolecule composed of amino acids and encoded by genes on DNA or RNA.

pseudoscience erroneous beliefs or practices falsely described as resulting from science.

replication the biochemical synthesis of new DNA molecules.

RNA ribonucleic acid; a nucleic acid resulting from transcription; can serve as the genetic material in some viruses.

samatha calm-abiding meditation.

samsara (Sanskrit and Pali) the material world into which beings are repeatedly reborn until enlightenment is achieved.

sangha a community of Buddhist monks, nuns, or other practitioners.

skandha commonly translated as aggregate; the five *skandhas* framework describes the lived experience through subject-object interaction.

sunyata commonly translated as emptiness; concept central to Madhyamaka philosophy and equated to *pratitysamutpada*.

sutra a Buddhist story or teaching.

Theravada one of two widely recognized branches of Buddhism (along with Mahayana) practiced in a variety of places such as Sri Lanka, Cambodia, Myanmar, and Thailand.

transcription the biochemical process of RNA synthesis from a DNA template molecule.

translation the biochemical process of protein synthesis from a RNA template molecule.

ultimate truth the ultimate nature of reality that can only be known through direct meditative experience, as expressed by *sunyata* in Madhyamaka philosophy.

Vajrayana considered by some to be a third major branch of Buddhism; also described as "tantric" or "esoteric" Buddhism; practiced in places such as Tibet, Mongolia, and Japan.

Vipassana insight meditation.

Zen school of Mahayana Buddhism that emphasizes meditation and mindful awareness; practiced in places such as China, Japan, and Vietnam.

Index